Kris Heyde

From Nucleons to the Atomic Nucleus

Perspectives in Nuclear Physics

With 106 Figures, 6 Tables,
11 Technical Boxes, and 1 Fold-out Chart

 Springer

Professor Dr. Kris Heyde
Vakgroep Subatomaire en Stralingsfysica
Institute for Theoretical Physics, University of Gent
Proeftuinstraat, 86
B-9000 Gent, Belgium
e-mail: kris.heyde@rug.ac.be

Cataloging-in-Publication Data applied for

Die Deutsche Bibliothek – CIP-Einheitsaufnahme

Heyde, Kris: From nucleons to the atomic nucleus: perspectives in nuclear physics; with 6 tables / K. Heyde. – Berlin; Heidelberg; New York; Barcelona; Budapest; Hong Kong; London; Milan; Paris; Santa Clara; Singapore; Tokyo: Springer, 1998
ISBN 3-540-63122-4

ISBN 3-540-63122-4 Springer-Verlag Berlin Heidelberg NewYork

© Springer-Verlag Berlin Heidelberg 1998
Printed in Germany

The use of general descriptive names, registered names, trademarks, etc. in this publication does not imply, even in the absence of a specific statement, that such names are exempt from the relevant protective laws and regulations and therefore free for general
use.

Typesetting: Data conversion by Springer-Verlag
Cover design: *design & production* GmbH, Heidelberg
SPIN 10546024   56/3144 – 5 4 3 2 1 0 – Printed on acid-free paper

To all the children
of the Fundación Niño Feliz from Bolivia
"de corazón a corazón"

Preface

The present text grew out of a number of lecture courses for advanced under-graduate and new graduate students in nuclear physics. They were given at summer schools in Leuven, Melbourne, and at study weeks for Dutch grad-uate students which aimed to emphasize fundamental and topical aspects of nuclear physics. On occasion, part of the present text was presented to stu-dents from a much wider field than just nuclear physics and also within a number of general physics colloquia, where, in addition to nuclear physicists, physicists from many other fields were present. In this respect, the intention is to present, in an amply illustrated form, the key questions that arise in nuclear physics. At the same time we try to show why a better understanding of the atomic nucleus is not only important in itself, but also yields essential insights into the many connections to other fields of physics. We thus concen-trate on the unifying themes rather than addressing in great detail particular subfields of nuclear physics.

The present project does not aim to be another comprehensive textbook on nuclear physics: Many of the detailed technical arguments that enter into the picture are not developed here as they would be in a more standard textbook. Instead they are presented using analogies, quite often with simple pictures and arguments that try to convey the general line of thinking and working in nuclear physics. The mathematics that we use is at the level com-monly employed to introduce the basic ideas of nuclear physics and quantum mechanics. On occasion, "Technical Boxes" are included, concentrating in some detail on specific issues that are important. Here, a number of theoreti-cal concepts as well as recent experimental techniques and efforts in advancing the present field of nuclear physics are illustrated. The way of citing refer-ences is also different from a standard textbook. No specific references are brought into the main text but, chapter by chapter, we list books, important review articles, some essential older papers, and also more popular discourses at the level of Scientific American, Physics Today, etc. for those wishing to pursue more detailed and in depth studies. The general level and, in partic-ular, the number and style of the illustrations is chosen in order to enable a broad readership to follow the major flow of the arguments. Thus it is our hope that readers will come to appreciate why the atomic nucleus is a vital 'laboratory' for investigating much of the working of the world around us and the kind of intriguing questions currently being studied experimentally and theoretically, both in small university research groups and by large-scale international collaborations. So, the text aims primarily at physics students but, in addition to physicists throughout the broader discipline of physics.

Much of the content, structure, and presentation has been influenced over the years by a large number of people with whom I had professional contact, but also by various scientists and friends who regularly asked me why the study of such tiny 'chunks' of matter could be so exciting. My long involvement in teaching has influenced the text on almost every page in its quest for clarity in answering the many questions concerning the 'big picture' of which nuclear physics is just a small (albeit essential) part. I would like to thank the physics students in Gent I taught over the last 15 years, as well as those at international schools and study weeks. The climate at the former Nuclear Physics Laboratory and the Laboratory for Theoretical Physics (now united into the 'Department of Subatomic and Radiation Physics') at the University of Gent has been and remains very stimulating and I have to thank in particular the many members (past and present) of the theory group in Gent: In alphabetical order I mention C. De Coster, B. Decroix, Y. De Wulf, N. Jachowicz, J. Jolie, L. Machenil, J. Moreau, S. Rombouts, J. Ryckebusch, L. Van Daele, M. Vanderhaeghen, V. Van der Sluys, P. Van Isacker, J. Van Maldeghem, D. Van Neck, W. Van Nespen, H. Vincx, G. Wenes and M. Waroquier. Also, the experimental groups in the department continuously reminded me that a description of the atomic nucleus, nice though it may look on paper, always has to pass the test of confrontation with experiment. The work carried out in our theory group has influenced and modified in almost every respect my view on how to study the atomic nucleus and why. I also appreciate very much the support I received about 30 years ago when studying at the Nuclear Physics Laboratory of the University of Utrecht, and in particular my supervisor P. Brussaard, whose influence was very decisive. Finally, I would like to thank A. Richter for his longstanding and inspiring experimental efforts to bring the best and most exciting out of nuclear physics. I value his deep sense of the questions that are important to address as well as his ability to extract from nuclear physics the aspects that are essential in the wider field of physics. His ideas, and in particular his recent article entitled *Trends in Nuclear Physics* [(1993) Nucl. Phys. A **553**, 417] have greatly influenced my own ideas.

Gent, August 1997   *Kris Heyde*

Contents

List of Technical Boxes

1. Introduction:
What Nuclear Physics is About

Nuclear physics, as the study of many-body fermion systems using both experimental and theoretical methods of research, is literally at the 'heart' of all matter.

All forms of matter we observe around us in the universe are built out of atoms of various types ranging from the lightest and most abundant hydrogen, through the light elements, to the heavy atoms and actinides. As a unifying feature, all atoms are formed by an atomic nucleus, which produces the Coulomb field that binds a number of electrons, thereby forming neutral atoms. Even though almost all the mass is concentrated in the central region, i.e., in the atomic nucleus with a size of 10^{-15} m, the scale determining the structure of ordinary matter is determined by the size of the atoms themselves, i.e., about 10^{-10} m. In a first, simple figure (Fig. 1.1), we illustrate this construction principle and see that it extends back to even smaller scales.

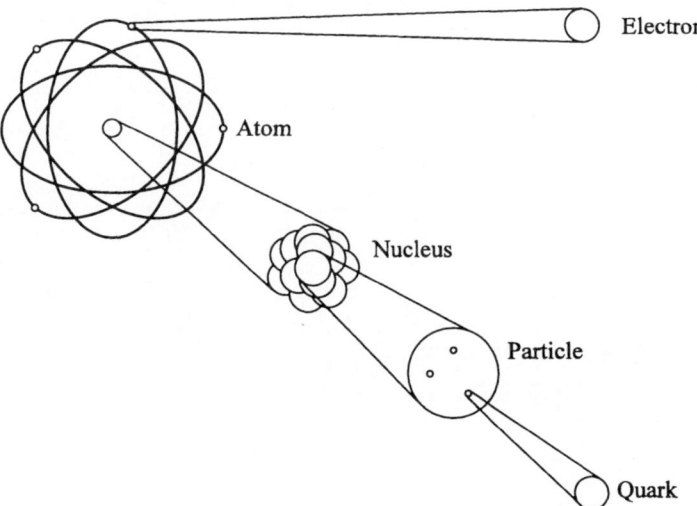

Fig. 1.1. The construction principles of matter, starting from the atomic scale (atomic nucleus and electrons) down to the quark structure, passing through the atomic nucleus with its proton and neutron constituents. (Taken from CERN/DOC ©1986, March, CERN Publication, with permission)

At present it is generally accepted that the nucleons constituting the nucleus are themselves built out of a few quarks constrained to form nucleons by the exchange of gluons. The picture of nucleons as built mainly from three quarks is only a rough approximation and, more generally speaking, the nucleon itself plays the role of a strongly interacting quantum chromodynamical many-body system. This interesting aspect, however, is largely beyond the scope of the present book.

Here, we shall concentrate on those aspects that yield a better insight into the way nucleons finally make up an atomic nucleus and give rise, through the nucleon–nucleon interaction inside the atomic nucleus, to a very rich pattern of phenomena at first sight not easy to predict or explain.

We shall begin by applying a magnifying glass to regular matter and try to observe at which stage the strong force binding the nucleons (protons and neutrons) becomes evident. The nucleons occupy a very tiny region of space and result in nuclear densities that are typically of the order of 10^{17} kg/m^3 and give rise to processes on an energy scale between 1 keV and, at the other extreme, 50–100 MeV. Thereafter we shall properly address the topic of particle-like properties. In Fig. 1.2, we show the typical density scale and energy scale at which atomic nuclei are located as compared to the phenomena of particle physics on one side and atomic, molecular, and solid-state physics and chemistry on the other side.

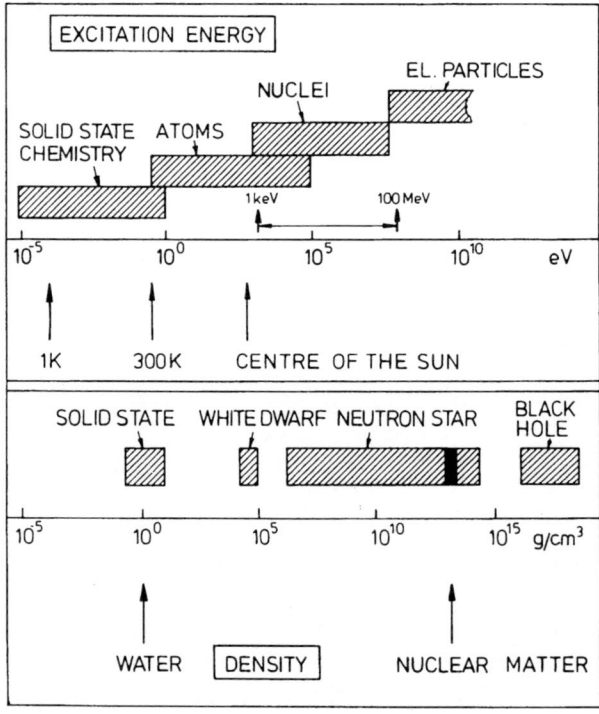

Fig. 1.2. Energy scale (*upper part*) situating the atomic nucleus with respect to the surrounding systems, e.g., the atomic, solid-state and chemical phenomena on one side and the domain of elementary particle physics on the other. The *lower part* shows the density scale, giving the position of the atomic nucleus with respect to regular matter, stellar structures, and black holes. (Taken from K. Heyde *Basic Concepts in Nuclear Physics* ©1994 IOP Publishing, with permission)

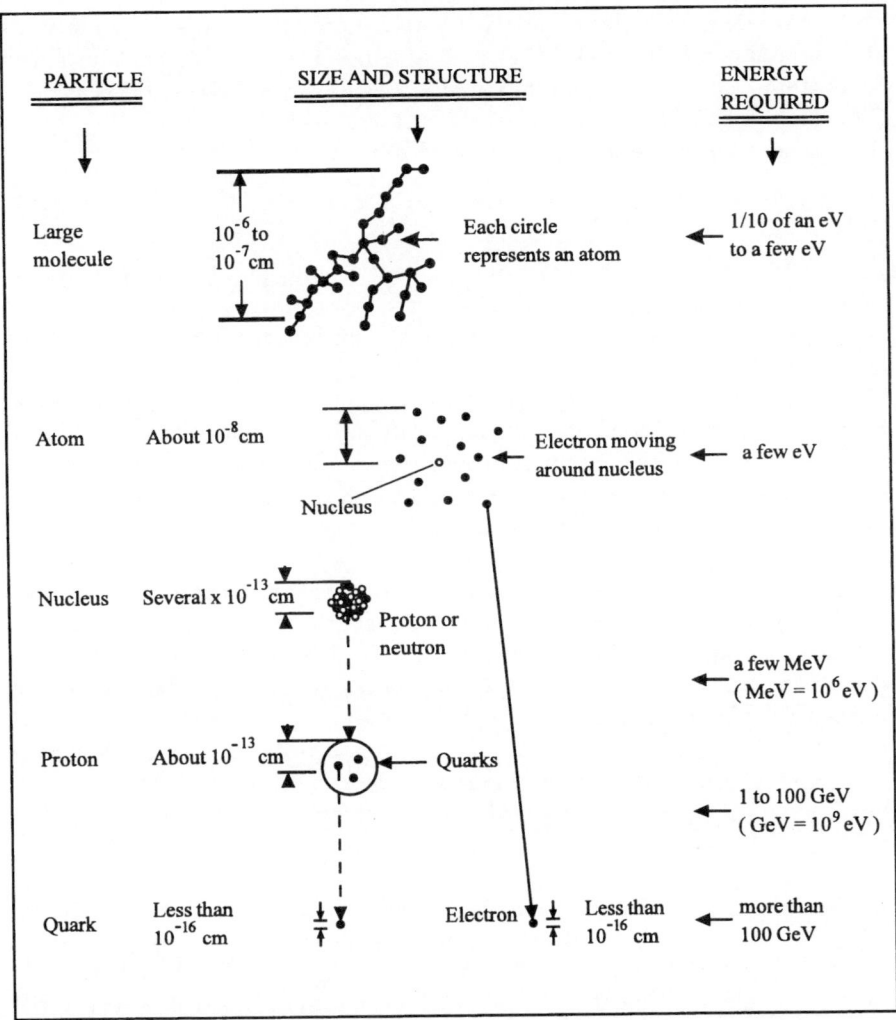

Fig. 1.3. Particle systems, their sizes and structures, and the energy required to probe them. We start from molecular structures and progress down towards the quark level

Descending in length scale from a typical large molecule, the road leading deep into matter by constantly increasing the resolving power of our microscope reveals a number of steps that are illustrated in Fig. 1.3. There are, however, a number of complications, due to the very process of trying to 'observe' ever smaller objects. The process of 'observing' with light (just a very tiny part of the full electromagnetic radiation spectrum characterized by wavelengths of 4–8×10^{-7} m) breaks down at atomic length scales so that large molecules form the limit of purely 'optical' exploration (Fig. 1.4a). Entering the domain of the atomic scale itself and going deeper, we move out of the classical regime and so have to take into account the quantum nature of matter. According to the de Broglie matter–wave duality, one can associate a wave-like character with any particle, with the wavelength (λ) and the momentum of the particle (p) being related by

$$\lambda = \frac{h}{p} . \tag{1.1}$$

This quantum character, which was shown to hold for the elementary constituents of matter such as electrons, neutrons, protons, etc., as demonstrated in the early interference experiments, has more recently been shown to hold for larger objects like atoms, too (Fig. 1.4b).

It has become clear that, descending to the level of the very small, getting inside the atomic nucleus and trying to observe not only the global atomic structure and density contours but also the way in which the nucleons themselves 'move' inside the nucleus and compose it, one needs to look on the quantum scale using probes accelerated to high enough energies that the wavelength is in the range of 10^{-14}–10^{-15}m.

It has been the tremendous technical advances in accelerators (the part producing high-energy and thus short wavelength probe), detectors (the part where the scattered particles are observed), and data analysis techniques (needed to make the real 'observation' process visible also to the human eye in the form of a scattering intensity distribution as a function of the scattering angle) for electrons, muons, and also hadronic probes that have made this 'magnifying' process possible. The result is a vast and rich laboratory for examining the special ways in which the interplay of nucleons, interacting through the nucleon–nucleon force, causes organization in the atomic nucleus.

Before entering our journey from nucleons to the more global properties of the nucleus, it is a good point to delineate the various main divisions of energy (and length) scale that can be made when observing the atomic nucleus (Fig. 1.5).

On the largest length scale, using lower energy projectiles, one 'sees' the nucleus as a whole and can observe a number of very interesting features that will be discussed more deeply in Chap. 3, i.e., the nuclear surface collective modes. The nuclear forces, discussed in Chap. 2, thereby reveal themselves through collective cooperative effects causing an almost uniform matter density to result with a sharp fall-off so that the nuclear radius becomes a well-defined quantity. Collective oscillations then become apparent, and, for nonspherical objects, collective rotations can even be generated.

Descending deeper inside, specific proton and neutron features become clearly 'visible' and so one now has to bridge the gap and consider how nucleons, interacting via the nucleon–nucleon force, account for collective phenomena. The use of an average or mean-field approximation and its understanding, when looking with not too high resolution, is clearly a major enterprise.

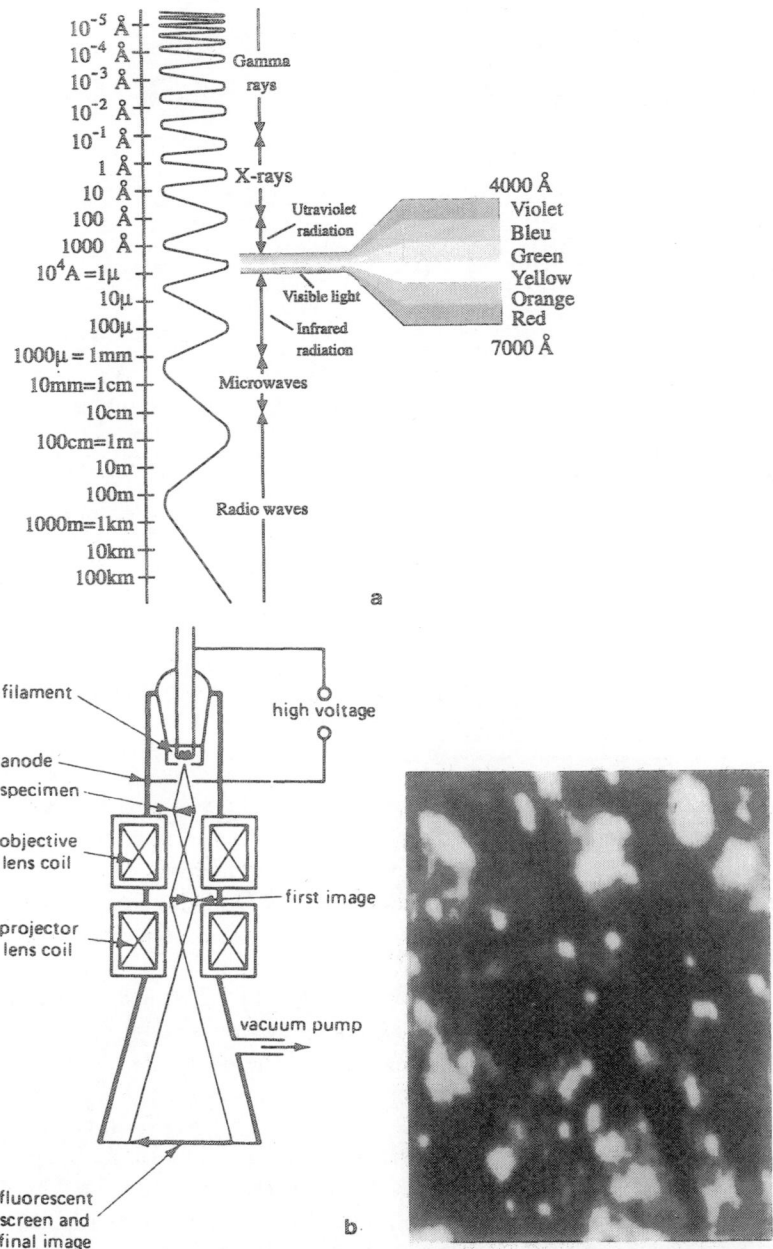

Fig. 1.4. (a) The electromagnetic spectrum extending from the shortest wavelength gamma rays to the very long wavelength radio waves. The visible part is magnified. (Taken from W. J. Kaufmann *Universe 3/E* ©1991 W. H. Freeman and Co., N.Y., with permission) (b) Schematic outline of an electron microscope enabling one to descend into the domain where molecular structures and large atoms become visible *(left-hand side)* (Taken from H. C. Ohanian *Modern Physics* ©1987 Prentice-Hall, N.Y., with permission) with (on the *right-hand side*) a micrograph of uranium atoms fixed on a thin film of carbon, taken with a high-power electron microscope at the University of Chicago. (By courtesy of A. V. Crewe, M. Utlaut, Univ. of Chicago) The magnification is about $\times 10^7$

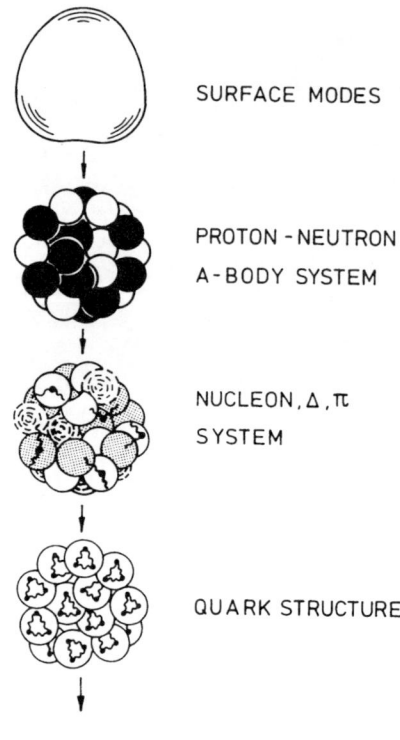

SURFACE MODES

PROTON - NEUTRON

A - BODY SYSTEM

NUCLEON, Δ, π

SYSTEM

QUARK STRUCTURE

Fig. 1.5. The length scales that characterize various levels of description of the atomic nucleus. In the *top part*, only superficial features are observable. Next, the specific proton and neutron excitations become "visible". At even smaller distances (higher energies) nucleon excitations (Δ, ...) and the pions mediating the nucleon force become observable. Finally, at the other extreme, quark degrees of freedom become accessible (*bottom part*). (Taken from K. Heyde *Basic Concepts in Nuclear Physics* ©1994 IOP Publishing, with permission)

Looking even more closely, one cannot proceed further with just nucleonic observables (protons, neutrons). The fact that nucleons are not the fundamental objects needed to fully understand both low- and higher-energy aspects is thus manifesting itself more and more clearly. On this scale, one can sometimes detect a nucleon itself in an 'excited' state and so cannot avoid bringing mesonic degrees of freedom into the picture.

At the other extreme of the energy scale, at very high energies, one eventually gets to grips with the complications of having an even deeper transition into a quark–gluon interacting system and requiring the full QCD theoretical framework. This, however, falls outside the scope of the present discussion.

In reading through the above introduction, one may get the impression of a very well 'layered' system where one can go from the global nuclear, collective properties through the nucleonic level and onwards to the scale where non-nucleonic degrees start becoming observable, and ending up with a fully particle-like formulation of nuclear physics. Instead of emphasizing this 'reductionistic' approach we point out that, at each level, a number of totally unexpected phenomena occur that could not easily, or not at all, be predicted in terms of just the deeper level constituents. Each 'level' very clearly shows its own fundamental characteristics and properties that are as basic as these on the other 'levels'. The theoretical approaches and experimental tools used to connect various levels brings in a deep degree of understanding of how to relate many-body properties to changing building blocks. As an example, one can try to describe the static average nuclear field starting from the basic nucleon–nucleon interaction, as observed between 'free' nucleons. This aim has not yet been fully accomplished and one has to bring into the process a number of simplifying approximations (how to go from nucleons interacting in 'free' space to nucleons interacting within the 'nuclear' medium; approximating the nucleon interactions by Hartree–Fock methods separating the mean-field components from residual interactions, etc.). Thus one sees that crossing borders and connecting the various length scales for describing the atomic nucleus is not a simple endeavor.

It would be a natural step now to discuss how the nucleons and their interactions give rise to both global bulk properties like nuclear radii, binding energy, etc., and local properties that are determined by the way the nucleons react to each other and give rise to an average field (Chap. 3). Let us, however, first stress that the study of the basic properties of the atomic nucleus, which in itself is a domain of science addressing questions of major importance, is strongly related (see Fig. 1.6) to other fields of science. Nuclear physics is very strongly interconnected to particle physics, astrophysics, astronomy, and cosmology. This simply reflects that the basic structures in the universe are themselves intimately linked to each other. Even though at the other extreme one encounters specialization into a large number of sub-domains in nuclear physics (this is very evident when one looks at a typical issue of the Physical Review or Nuclear Physics, to quote just two journals, and observes the many divisions and subdivisions), the unity behind all experimental and theoretical research cannot be stressed strongly enough. It is one of the aims of the present book to show, by looking at the various length scales in the nucleus and the rich spectrum of phenomena characterizing them, that the unifying aspects dominate over the subdivisions and to point towards a number of general principles describing how the nucleons move inside, and at the same time constitute the entity representing the atomic nucleus.

At the end of each chapter, we give references to specific textbooks, articles and more popular texts. These works cover quite general aspects of that particular chapter and will be grouped together and put in some perspective with a few descriptive sentences.

At the end of this introductory chapter, we list a number of nuclear physics textbooks concentrating on the whole spectrum of phenomena. We also refer to a number of works discussing general trends in nuclear physics as well as some of the recent long-term plans.

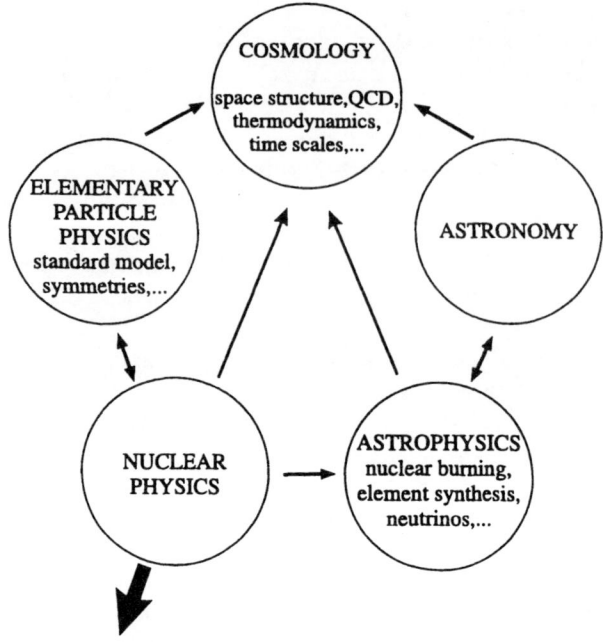

Fig. 1.6. Diagram illustrating various connections between nuclear physics and other fields of physics. (Taken from A. Richter (1993) Nucl. Phys. A, **553**, 417c. Elsevier Science, NL, with kind permission)

1.1 Further Reading

Nuclear Physics, Intermediate Energy Physics,
Introductory Particle Physics

For this introductory chapter, we first list a set of textbooks that discuss the global field of nuclear physics, and where both the experimental developments and theoretical methods are presented. Some of these books cover both nuclear and particle physics in a consistent way.

1.1 Blatt J.M., Weisskopf V.F. (1952) *Theoretical Nuclear Physics* (Wiley, New York)
1.2 Bohr A., Mottelson B. (1969) *Nuclear Structure* Vol. 1 (Benjamin, New York); (1975) *Nuclear Structure* Vol. 2 (Benjamin, New York)
1.3 Evans R.D. (1955) *The Atomic Nucleus* (McGraw-Hill, New York)
1.4 Fermi E. (1950) *Course on Nuclear Physics* (University of Chicago Press, Chicago)

1.5 Frauenfelder H., Henley E.M. (1991) *Subatomic Physics* 2nd edn. (Prentice-Hall, New York)

1.6 Green A.E.S (1955) *Nuclear Physics* (McGraw-Hill, New York)

1.7 Jelley N.M. (1990) *Fundamentals of Nuclear Physics* (Cambridge University Press, Cambridge)

1.8 Kaplan I. (1963) *Nuclear Physics* 2nd edn. (Addison-Wesley, Reading, MA)

1.9 Krane K.S. (1987) *Introductory Nuclear Physics* (Wiley, New York)

1.10 Marmier P., Sheldon E. (1969) *Physics of Nuclei and Particles* Vol. 1 (Academic, New York)

1.11 Meyerhof W. (1967) *Elements of Nuclear Physics* (McGraw-Hill, New York)

1.12 Preston M.A., Bhaduri R.K. (1975) *Structure of the Nucleus* (Addison-Wesley, Reading, MA)

1.13 Rose M.E. (1965) α, β and γ *Spectroscopy* Vol. 1 ed. by K. Siegbahn (North-Holland, Amsterdam)

1.14 Segré E. (1982) *Nuclei and Particles* 3rd edn. (Benjamin, New York)

1.15 Valentin L. (1981) *Subatomic Physics: Nuclei and Particles* Vols. 1 and 2 (North-Holland, Amsterdam)

1.16 Williams W.S.C. (1991) *Nuclear and Particle Physics* (Oxford University Press, Oxford)

1.17 Wong S.S.M. (1990) *Introductory Nuclear Physics* (Prentice-Hall, New York)

General Textbooks Discussing Related Topics

We also mention just a few general physics textbooks that can be consulted since the subfield of nuclear physics is not separate from the rest of physics and quite often concepts from the more "classical" physics domain are needed. In particular, a few modern texts bridging the gap between the classical physics curriculum and what is known as "modern physics" (topics developed in the last 30–40 years) are included here.

1.18 Alonso M., Finn E. (1971) *Fundamental University Physics* Vol. II and Vol. III (Addison-Wesley, Reading, MA)

1.19 Davies P. (ed.) (1989) *The New Physics* (Cambridge University Press, Cambridge)

1.20 Feynmann R.P., Leighton R.B., Sands M. (1965) *The Feynmann Lectures on Physics* Vol. III (Addison-Wesley, Reading, MA)

1.21 Krane K.S. (1996) *Modern Physics* 2nd edn. (Wiley, New York)

1.22 Lorrain P., Corson D.R. (1978) *Electromagnetism : Principles and Application* (Freeman, New York)

1.23 Ohanian H.C. (1987) *Modern Physics* (Prentice Hall, New York)

1.24 Orear J. (1979) *Physics* (Collier MacMillan International Editions, New York)

1.25 Panofsky W.K.H., Phillips M. (1962) *Classical Electricity and Magnetism* 2nd edn. (Addison-Wesley, Reading, MA)
1.26 Rohlf J.W. (1994) *Modern Physics from α to Z* (Wiley, New York)

Quantum Mechanics Texts

A couple of basic quantum mechanics texts that can be consulted when more explicit issues of quantum mechanics show up or when reference is made to the importance and impact of quantum mechanics on nuclear physics:

1.27 Flügge S. (1974) *Practical Quantum Mechanics* (Springer, New York)
1.28 Greiner, W. (1997) *Quantum Mechanics – Special Chapters* (Springer, Berlin Heidelberg)
1.29 Merzbacher E. (1970) *Quantum Mechanics* 2nd edn. (Wiley, New York)
1.30 Sakurai J.J. (1973) *Advanced Quantum Mechanics* (Addison-Wesley, Reading, MA)
1.31 Schiff L.I. (1968) *Quantum Mechanics* (McGraw-Hill, New York)

Mathematical References

It may well be interesting on occasion to have some guidance with respect to mathematical methods and techniques, in particular oriented towards looking up properties of solutions of the Schrodinger equation, studying eigenvalue problems and associated matrix manipulations. A short list is given.

1.32 Abramowitz M., Stegun J.A. (1964) *Handbook of Mathematical Functions* (Dover, New York)
1.33 Arfken G. (1985) *Mathematical Methods for Physicists* 3rd edn. (Academic, New York)
1.34 Wilkinson J.H. (1965) *The Algebraic Eigenvalue Problem* (Clarendon, Oxford)

The aim of the reference lists given here and in other chapters is not to be exhaustive, but to provide essential further reading at various levels: popular, intermediate, textbooks concentrating on specific issues, and finally advanced technical reviews.

2. Nucleons in Interaction: The Nucleon–Nucleon Force

2.1 Introduction

Nucleons moving inside the nucleus are subject to the strong force even though the nucleon–nucleon interactions are greatly modified with respect to the interaction between free nucleons. The interaction inside the nucleus is dependent on the precise structure of the nuclear medium and is strongly state and mass dependent. In Chap. 3 we shall discuss in more detail how the 'free' nucleon–nucleon interaction is modified when the two interacting nucleons are brought inside an atomic nucleus with a total of A nucleons, and, moreover, when they are moving in specific single-particle orbitals in a spherical average field.

We shall discuss the nucleon–nucleon force structure mainly from a phenomenological point of view and will use symmetry constraints to learn a number of robust characteristics of the nucleon–nucleon interaction. We also illustrate how one can pin down the nuclear interaction starting from scattering experiments between free nucleons. In the final section, we look at the potentials describing the interaction as a function of the separation between the two nucleons and eventually make contact with a description that uses the quark degrees of freedom explicitly, albeit in a pedestrian way.

2.2 The Symmetries of the Nucleon–Nucleon Force

Nucleons can be depicted as interacting via essentially two-body forces that are determined mainly by the separation of the two nucleons. Inside an atomic nucleus, unlike the interactions felt by electrons in an atom, i.e., the one-body Coulomb field to which specific two-electron interactions can be added, there is no natural one-body potential acting on the nucleons. One can, however, use Hartree–Fock theory to derive an "effective" mean field which mainly represents the force experienced by a given nucleon due to its average interaction with the remaining $A - 1$ nucleons. This will be discussed in more detail in Chap. 3.

The nucleus which exhibits the two-body nucleon interaction most clearly is the deuteron, but the information that can be deduced from it is rather limited because this system has only one bound state. Rather than trying to extract information about the nucleon–nucleon force by studying the bound state properties of light nuclei, we examine what symmetry constraints can teach us about the nucleon–nucleon force.

One very important concept is that of charge symmetry: This means that, ignoring the Coulomb force, changing protons into neutrons and vice versa does not modify the nuclear properties. This symmetry is derived from the study of various properties relating to energy spectra in mirror nuclei. The symmetry can even be extended and it reveals that the nucleon–nucleon inter-action is charge independent for those states that can be realized (a proton–neutron system can form particular states that are forbidden in the proton–proton or neutron–neutron system), and thus p–p, p–n and n–n forces all give the same contribution to the binding energy and other properties of nuclei.

The nucleons in the nucleus, and also free nucleons, are described by their spatial coordinates, their linear and angular (spin) momenta, and various other intrinsic properties (electric charge, magnetic moment, etc.). A number of invariances imply symmetry constraints on the radial dependence of the force and on the precise combination of the nucleon variables that appear when expressing the nucleon–nucleon force.

Invariance under a translation of the two interacting nucleons implies that the interaction depends only on the relative separation $r = r_1 - r_2$. The linear momentum also appears only as the relative momentum because the center of mass will not be a variable affecting the nucleon interactions, i.e., only $p = 1/2(p_1 - p_2)$ occurs. Further constraints are that the nucleon–nucleon force should not change under a rotation of the coordinate system, should be time-reversal invariant, should conserve parity (for the strong force), and should possess permutation symmetry upon the exchange of the two inter-acting nucleons. Including, besides the relative coordinate and the relative momentum, the intrinsic spin, isospin, total spin, and total angular momen-tum, it can be shown that the most general form of the interaction potential can be expressed as

$$
\begin{aligned}
V\left(r; \sigma_1, \sigma_2; \tau_1, \tau_2\right) = \; & V(r) + V_\sigma(r)\sigma_1 \cdot \sigma_2 + V_\tau(r)\tau_1 \cdot \tau_2 \\
& + V_{\sigma\tau}(r)\left(\sigma_1 \cdot \sigma_2\right)\left(\tau_1 \cdot \tau_2\right) + V_{LS}(r)L \cdot S \\
& + V'_{LS}(r)\left(L \cdot S\right)\left(\tau_1 \cdot \tau_2\right) + V_T(r)S_{12} \\
& + V'_T(r)\,S_{12}\tau_1 \cdot \tau_2 + V_Q(r)Q_{12} \\
& + V'_Q(r)Q_{12}\tau_1 \cdot \tau_2 + V_{\sigma p}(r)\left(\sigma_1 \cdot p\right)\left(\sigma_2 \cdot p\right) \\
& + V'_{\sigma p}(r)\left(\sigma_1 \cdot p\right)\left(\sigma_2 \cdot p\right)\left(\tau_1 \cdot \tau_2\right) \quad\quad\quad (2.1)
\end{aligned}
$$

with $\sigma_1, \sigma_2, \tau_1, \tau_2, S, L$ denoting the intrinsic spin, isospin, and the total spin and total orbital angular momentum operators, respectively.

The terms appearing here are a two-body spin–orbit force given by

$$
L \cdot S = (l_1 + l_2) \cdot \frac{1}{2}(s_1 + s_2)\,, \quad\quad\quad (2.2)
$$

a tensor force (which closely ressembles the interaction between two magnetic dipole moments placed at a relative distance r) S_{12}

$$S_{12} = \frac{3}{r^2} \left(\boldsymbol{\sigma}_1 \cdot \boldsymbol{r}\right) \left(\boldsymbol{\sigma}_2 \cdot \boldsymbol{r}\right) - \boldsymbol{\sigma}_1 \cdot \boldsymbol{\sigma}_2 , \tag{2.3}$$

and the quadratic spin–orbit operator

$$Q_{12} = \frac{1}{2} \left[\left(\boldsymbol{\sigma}_1 \cdot \boldsymbol{L}\right) \left(\boldsymbol{\sigma}_2 \cdot \boldsymbol{L}\right) + \boldsymbol{\sigma}_2 \cdot \boldsymbol{L} + \boldsymbol{\sigma}_1 \cdot \boldsymbol{L}\right] . \tag{2.4}$$

The radial dependence is at present undetermined but here, too, important constraints exist if we insist that, for nucleon separations of 1.5–2.0 fm, the potential should be of a Yukawa form due to the one-meson exchange underlying the nucleon–nucleon force. We shall come back to the radial dependence in more detail in Sect. 2.4.

2.3 Experimental Knowledge of Nucleon–Nucleon Interactions

The specific but rather general form of the nucleon–nucleon force has to be determined in such a way that it correctly describes the scattering between free nucleons, and this as a function of the bombarding energy. It should also provide a good description of the polarization effects that appear when particles with intrinsic spin scatter off each other.

The easiest experiments involve scattering protons off protons (p–p) because proton beams can be readily produced at varying energies at accelerator facilities with cyclotrons. Scattering of neutrons from protons is much more difficult. Producing low energy neutrons is relatively easy and they are copiously produced at nuclear reactors; but higher energy neutrons have to be generated via specifically chosen nuclear reactions. Here, the intensity never becomes very high. One can, in principle, also perform n–n scattering experiments, but in most situations one needs subtraction measurements since pure neutron targets cannot be made. In the case of p–p scattering, the Coulomb effect needs to be isolated from the nuclear part but this is not difficult.

Scattering experiments over a large range of bombarding energies have been carried out over the years and constitute an extensive data base against which the nucleon–nucleon force can be tested. The procedure used is as follows: Starting from a given choice of the nucleon–nucleon force, the nuclear phase shifts for the various partial waves describing the colliding nucleons are described theoretically as well as possible. A least-squares fitting procedure can then fit the free nucleon–nucleon interaction in quite a realistic way. We shall not go into detail but the basic ingredients result from solving the Schrödinger equation for positive energies using as the potential the nucleon–nucleon potential of (2.1), and constraining the asymptotic wave function to be the superposition of an incoming plane wave and a scattered outgoing spherical wave. A partial wave decomposition of the full wave function then allows one to determine, for each individual angular momentum component, the phase shift with respect to the unperturbed plane wave phases. Details can be found in quantum mechanics texts on scattering processes. We illustrate in Fig. 2.1 a set of such phase shifts for the particular case of $T = 1$ scattering (pp, nn, or pn scattering) for relative angular momenta with the value 0 (S-wave scattering), 1 (P-wave scattering) and 2 (D-wave scattering). Consistent with the Pauli principle, in the S and D cases, the total intrinsic spin equals 0 whereas for the P-wave the total instrinsic spin becomes 1 and thus three different P-scattering states can be realized. A few of the most common forms of interaction used to describe the nucleon–nucleon force in a realistic way are the Hamada-Johnston hard-core potential, the Reid soft-core potential, and the Tabakin potential (see [2.5], Chap.3).

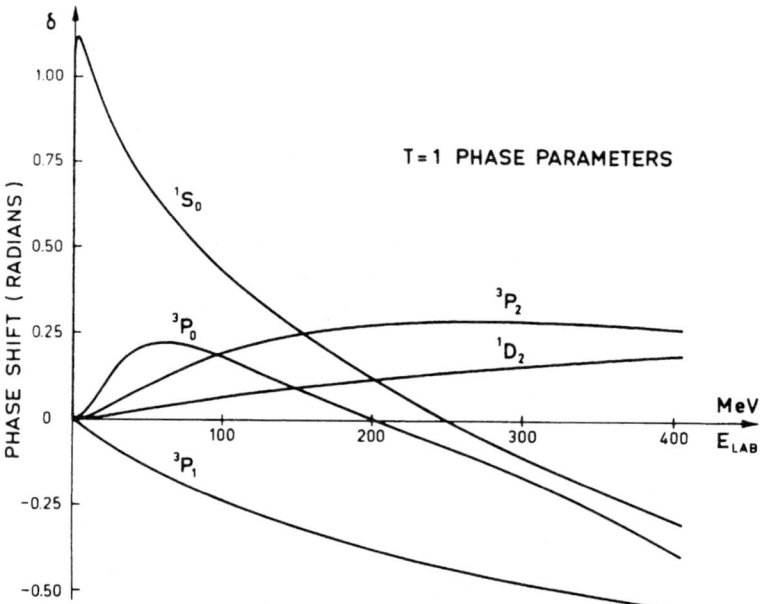

Fig. 2.1. The nuclear phase shifts δ for the scattering of nucleons in the isospin $T = 1$ channel (pp, nn, and np scattering) and this for motion with the relative angular momentum $\ell = 0, 1, 2$ (S, P and D states) and for spin-singlet and spin-triplet states. The energy extends up to a laboratory energy of $E_{\mathrm{lab}} \simeq 400$ MeV. (Taken from A. Bohr, B. Mottelson *Nuclear Structure* Vol. 1 (Fig. 2.24, p.264) ©1969 W.A. Benjamin. Reprinted with permission from Addison-Wesley, Longman Inc.)

The detailed analysis is further complicated, compared to the textbook case of potential scattering, purely by the appearance of the spin–orbit, spin–orbit force and tensor terms that couple various spin–orbital–isospin channels.

The phase shifts are not enough to determine the general force fully and, besides, a number of spin polarization observables need to be determined. In the scattering of two spin-1/2 particles one has 16 polarization observables, which, however, are not all independent of one another. One of the major variables is called the analyzing power. It is obtained when polarized particles are scattered from an unpolarized target and the polarization of the particles in the final state is not detected. This analysing power is only one of the five independent polarization variables that can be measured (see [2.5], Chap.3).

If the energy in the scattering between nucleons becomes high enough, highly inelastic processes can occur with a number of mesons being produced. With increasing energy, a large number of inelastic processes can start to occur in which one or both of the nucleons become excited internally. Generally, up to 300 MeV scattering energy, the phase shifts are essentially real quantities but at much higher energies, both real and imaginary components appear signaling the opening of inelastic channels. Full lists of phase shifts up to 1 GeV are available in the literature (see e.g. Arndt, Hyslop and Roper).

In low-energy scattering, where the relative angular momentum describing the interacting nucleons is mainly $l = 0$, properties other than the phase shift can be used to fix the nucleon–nucleon force. The most important quantities in this regime are the 'scattering length' and the 'effective range'. They are partly related to the deuteron binding energy and can be obtained, e.g., from the scattering of slow neutrons off protons in H_2 molecules. We shall not go into these details either but they can be found in quantum mechanics texts concentrating on the scattering regime (positive energy states) in potential problems (see further reading suggestions in Chap. 1).

2.4 The Nucleon–Nucleon Force

2.4.1 Potential Models

In trying to determine the radial shape of the strength functions describing the nucleon–nucleon force as given in (2.1), one is well advised to follow Yukawa and assume that the strong force coupling two nucleons is mediated mainly by mesons. In that picture, it is the meson–nucleon coupling that is responsible for the actual magnitude of the nucleon–nucleon force. So Yukawa forms can be a good starting point for understanding the potential shape describing the nucleon–nucleon force at separations of 1–2 fm.

The simplest case is that in which the force is due entirely to the exchange of a single pion (one-pion exchange potential, OPEP): the mass is almost 140 MeV and thus corresponds to a length scale of 1.4 fm. At shorter distances, more than one pion could be exchanged or the exchange of heavier mesons may become more important. The shorter range behavior is clearly manifest in the $^1S_0$ scattering channel: around 250 MeV laboratory energy, the scattering phase changes sign indicating that the radial force changes from attractive to repulsive. Such a repulsive character, which is indicative of a hard (or soft) core in the nucleon interaction energy when the distance is of the order of 1 fm, can be expected on the basis of the quark content of the composite nucleons. The two times three 'constituent' quarks which are in the lowest possible state for the isolated nucleon, cannot stay in this state when nucleons start to 'overlap' spatially. Three of the six quarks have to be promoted into higher energy states and this will cost a large amount of energy, expressed by the repulsive character of the potential describing the nucleon–nucleon force. This highly qualitative picture has not been much improved upon, even today: The energy is very low compared to the energy of quantumchromodynamics (QCD) processes and pertains to the 'non-perturbative' regime of QCD.

The present-day picture of the nucleon–nucleon interaction is depicted in Fig. 2.2 where a number of distinct distance scales are observed. The longest range properties are determined fully by the one-pion exchange process. It is natural then that the exchange of heavier mesons, or more pions, will give rise to the radial behavior in the region between 1 and 2 fm. And, finally, near 1 fm and at smaller separations, the hard core starts to play the dominant role.

In practice, in a hadron–meson coupling theory, problems also appear because the pion–nucleon coupling strength obtained from experimental data cannot be inserted in the pion–nucleon vertex inside the nucleus. Medium modifications change this but are very difficult to calculate; so the coupling constant is generally taken as an adjustable parameter.

The approach we have used to determine the nucleon–nucleon force over these three distance intervals is based on meson exchange processes: it is actually a phenomenological approach. A major drawback is the set of coupling strengths between the exchanged mesons and the nucleon as well as the appearance of the hard core, which mainly reflects our lack of understanding of

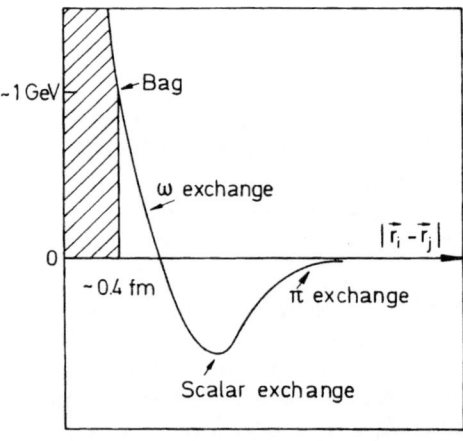

Fig. 2.2. Schematic drawing of the nucleon–nucleon potential (radial behavior). Three regions are distinguished: (i) beyond 1.5–2 fm, the one-pion exchange part (OPEP) dominates, (ii) in the intermediate region, heavier scalar mesons are exchanged and, (iii) at the smallest separations, a repulsive hard core shows up (Adapted from G.E. Brown, M. Rho (1983) *Physics Today* Feb. 24)

the processes at these small separations. In most low-energy processes that will be discussed later on we can nonetheless use this method with great success.

A different approach is one which incorporates all our detailed knowledge about hadrons and puts in the very short range properties, on which we have only limited knowledge, in a phenomenological way. The two examples of this approach are the Paris and Bonn potentials that have been and still are widely used. The two groups have essentially assumed the same behavior at larger separations but have used different techniques to describe the short-range characteristics. One of the differences is in the treatment of nucleon–antinucleon interactions, which have to be described in a relativistic theory. However, if one compares the phase shifts calculated for the most important nucleon–nucleon scattering channels with the data, the agreement is impressive for both potentials.

2.4.2 Relation to Quark Model

Even though in almost all practical applications in the nuclear physics domain the nucleon–nucleon interaction can be treated either in a phenomenological way (making use of the potentials describing the mutual interaction), or in a more sophisticated way using the modern Paris or Bonn potentials, one remains within a framework where, at most, the meson degrees of freedom enter the picture, albeit in a procedure which determines the effective nucleon–nucleon force strength. However, physicists are convinced that, in reality, the strong force that acts between nucleons is only a remnant of the interactions existing on a higher energy scale, namely, in the realm of QCD.

In order to describe nucleon–nucleon interactions it is desirable to study interaction processes amongst at least six quarks. There are a number of constraints though. If nucleons are well separated, the quark description needs to imply that the six quarks condense into two groups of three as the individual nucleons and that the interaction derived from the quark description should be in agreement with the OPEP picture discussed before (Fig. 2.3). Also, the intermediate distance scale where more and/or heavier bosons are exchanged should also be present in the quark model. At present, quite some work has gone into understanding this regime and a low-energy theory (the nuclear physics domain) is beginning to be constructed out of the rules governing QCD. The solution to this problem is hampered because it needs a solution of the equations of QCD in a regime that cannot be treated using perturbation theory, and where, at present, no exact solutions are possible.

One can shed light on the process intuitively by comparing it with the way forces between neutral molecules arise out of the Coulomb force, leading to the condensation of molecules in a liquid or solid phase. This is the van der Waals force and it provides an analogy with the way that two 'bags' of quarks each bound into a nucleon may still feel some remnant of the quark interactions. We shall not build too far on the analogy but the argument

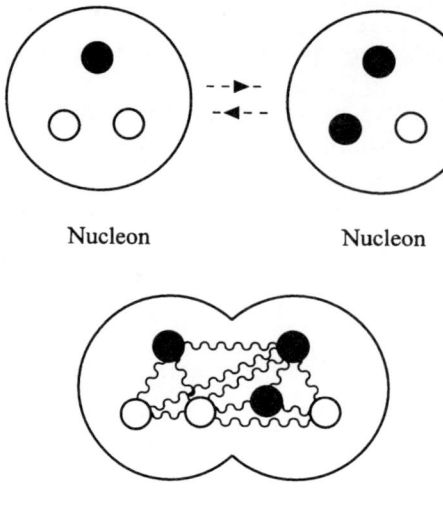

Nucleon   Nucleon

Di - nucleon

Fig. 2.3. The nucleon–nucleon interactions, viewed schematically, on the level of the constituent quarks. For large separations, meson exchange results as a van der Waals residual part. For very short distances, a six-quark system results where a clear-cut separation into two nucleons is no longer possible. Gluon exchange is drawn as a wavy line in the latter part

runs as follows: Shape fluctuations in a molecule will cause the otherwise spherical charge distribution to acquire a finite electric dipole moment. The electric field generated in this way can induce an electric dipole moment in the other molecule through its polarizing action. The net effect is a dipole–dipole interaction which is always attractive and represents a force with a radial dependence which goes like r^{-7}. The analogy with nucleons and quarks can then be formulated in a straightforward way with molecules now being replaced by nucleons containing quarks with a net vanishing 'color' charge. Although nucleons are 'neutral' with respect to their quark content, QCD theory yields interaction contributions between quarks in spatially separated nucleons that result in a residual color van der Waals force. The explicit structure of the force resembles the tensor interaction form given in (2.3) and is called the Breit interaction between quarks. A number of problems arise if one attempts to extend this rather naïve idea too far in particular when it comes to understanding the range of the nucleon–nucleon force as compared to the experimental nucleon–nucleon force radial structure. Various approaches are currently being pursued to better understand this intermediate region between nuclear and particle physics (e.g., chiral perturbation theory, numerical lattice QCD studies, low-energy theorems derived from QCD).

2.5 How Do Nucleons Interact Inside the Nucleus? The Role of the Nuclear Medium

A detailed knowledge of the nucleon–nucleon interaction is a very important step in understanding how those nucleons will interact pairwise when inside an atomic nucleus where many other nucleons are present and occupy well-defined quantum states. This environment will modify the 'bare' or 'free' nucleon–nucleon force in an important way into an 'effective' two-body nucleon force. The nucleon–nucleon force problem immediately becomes an A-body problem, where A is the number of nucleons in the nucleus. A large amount of work has gone into solving this problem and some of the pioneering is due to Scott-Moszkowski who devised a separation method which allows a type of cancellation between the internal hard-core region and the highly attractive intermediate distance scale component in the interaction leaving only a slightly attractive and smooth tail (the tail of the OPEP) at larger nucleon separations. Kuo and Brown have studied ways to correct for the presence of the medium by actually calculating the 'effective' force. In the next chapter these procedures will be illustrated and their consequences discussed.

In Chap. 3 we concentrate on the details of the nuclear structure. We will notice that in most cases the nucleon–nucleon interaction determines nuclear properties by the virtue of the effective force acting in a given mass region with nucleons moving in specific single-particle orbitals. In many cases one does not need the specific form of the interaction potential itself. The radial integrals and the angular momentum to which the nucleons are coupled are major factors in determining the nuclear properties. The nuclear matrix elements can be found by fitting to the large body of experimental data. This method will also be discussed in Chap. 3.

2.6 Further Reading

First we list some textbooks that give extensive coverage to or almost uniquely concentrate on nuclear interactions and the nucleon–nucleon force

2.1 Bhadhuri R.K. (1988) *Models of the Nucleon* (Addison-Wesley, Reading, MA)

2.2 Brown G.E., Jackson A.D. (1976) *The Nucleon–Nucleon Interaction* (North-Holland, Amsterdam)

2.3 Brown G.E. (1964) *Unified Theory of Nuclear Models* (North-Holland, Amsterdam)

2.4 Siemens P.J., Jensen A.S. (1987) *Elements of Nuclei: Many-body Physics with the Strong Interaction* (Addison-Wesley, Reading, MA)

2.5 Wong S.S.M. (1990) *Introductory Nuclear Physics* (Prentice Hall, New York)

A popular description of nuclear forces is given by

2.6 Beurty R., Saudinos J. (1990) Des forces nucléaires classiques à l'interaction forte élémentaires, Clefs CEA, n° 18, 2

The original article describing the nuclear force as conceived by Yukawa is

2.7 Yukawa H., (1935) Proc. Phys. Soc. Japan **17**, 48

Forms of the nucleon–nucleon interaction, the relation to nuclear phase shifts, and some of the most popular realistic nucleon–nucleon potentials are discussed in

2.8 Arndt R.A., Hyslop III J.S., Roper L.D. (1987) Phys. Rev. **D35**, 128
2.9 Hamada T., Johnston I. (1962) Nucl. Phys. **34**, 382
2.10 Machleidt R., Holinde K., Elster Ch. (1978) Phys. Rep. **149**, 1
2.11 Okubo S., Marshak R.E. (1958) Ann. Phys. **4**, 166
2.12 Tabakin, F. (1964) Ann. Phys. (N.Y.) **30**, 51
2.13 Vin Mau R. (1979) in: *Mesons in Nuclei*, eds. Rho M. and Wilkinson D.H. (North-Holland, Amsterdam)

In considering the quark content of nucleons, the description of the nucleon–nucleon interaction becomes significantly modified. Some important textbooks are

2.14 Close F.E. (1979) *An Introduction to Quarks and Partons* (Academic, New York)
2.15 Gottfried K., Weisskopf V.F. (1984) *Concepts of Particle Physics* Vol. I and (1986) Vol. II (Oxford University Press, New York)
2.16 Halzen F., Martin A.D. (1984) *Quarks and Leptons* (Wiley, New York)

Three articles concentrating on bag models of hadrons are

2.17 Brown G.E., Rho M. (1983) Physics Today **36**, 24
2.18 De Tar G.E., Donoghue J.F. (1983) Ann. Rev. Nucl. Part. Sci. **33**, 235
2.19 Thomas A. (1983) Adv. Nucl. Phys. **13**, 1

Discussions on how the nucleon–nucleon interaction should be modified in order to act as an effective force in the nuclear medium:

2.20 Kuo T.T.S., Brown G.E. (1966) Nucl. Phys. **85**, 40
2.21 Kuo T.T.S. (1974) Ann. Rev. Nucl. Sci. **24**, 101
2.22 Scott B.L., Moszkowski S.A. (1961) Ann. Phys. (NY) **14**, 107
2.23 Scott B.L., Moszkowski S.A. (1962) Nucl. Phys. **29**, 665

3. Introducing the Atomic Nucleus: Nuclear Structure

3.1 Bulk or Global Properties of the Atomic Nucleus

Here we consider the large-scale structure of the atomic nucleus. A nucleus comprises a system of A nucleons, which, when well separated from each other and outside the range of the strongly attractive nuclear force, has as its energy just the sum of the rest mass of the individual nucleons (Fig. 3.1 left). In bringing, in a hypothetical way, the nucleons close together, at the range of the strong force (approximately 10^{-14}–10^{-13}m), condensation into the bound atomic nucleus will result (Fig. 3.1 right) with a release of the corresponding binding energy. This causes an observable deficit in the mass of the actual nucleus as compared to the sum of the masses of the individual nucleons. This "loss" of mass is known as a mass defect.

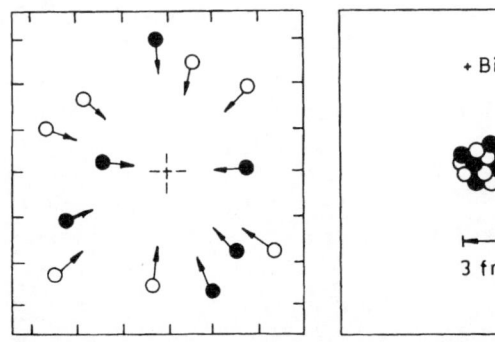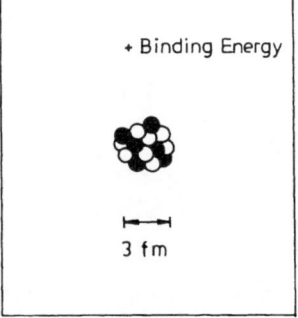

Fig. 3.1. Schematic illustration of the 'condensation' process in which, under the influence of the nucleon–nucleon strong short-range binding forces, a bound nucleus is formed and the corresponding binding energy is released resulting in a mass defect. The nuclear size is about 3 fm for a light nucleus. (Taken from K. Heyde *Basic Concepts in Nuclear Physic* ©1994 IOP Publishing, with permission)

The salient features of the so-formed 'drop' of nuclear matter, which in a first (static) approximation may be described using a liquid drop model, can, with the Hartree–Fock approach, be connected to the underlying nucleon–

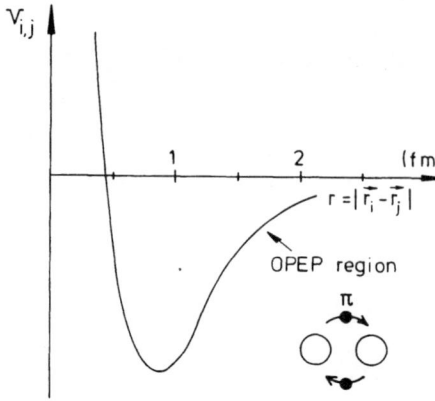

Fig. 3.2. Illustration of the short-range repulsive part and the long-range OPEP(one-pion exchange potential) tail of the nucleon–nucleon interaction. The one-pion exchange is shown schematically

nucleon force acting amongst the individual nucleons (see Fig. 3.2 for a simple picture of the n-n force).

The most important and fundamental of all elementary modes describing the atomic nucleus assumes the existence of an average one-body field, at least for describing low-energy phenomena and even when including the residual nucleon–nucleon interaction. This holds the nucleons together in a self-consistent way within the average field that they create.

To a very good approximation, each of the A nucleons moves in an almost independent way in a single-particle field $U(\boldsymbol{r}_i)$. The correctness of this assumption is the key to the shell model and its many accomplishments. Its successful application throughout the whole nuclear mass table will be discussed in the next section. Self-consistent Hartree–Fock calculations have given a firm basis to the independent-particle shell model of the nucleus. The average field and the corresponding single-particle wave functions and single-particle energies, starting from the kinetic energies of the individual nucleons and their two-body interactions, described by the potential $V(\boldsymbol{r}_i, \boldsymbol{r}_j)$, are obtained using the well-known (Brueckner)–Hartree–Fock method as follows:

One starts from the many-body Hamiltonian

$$H = \sum_{i=1}^{A} t_i + \frac{1}{2} \sum_{i,j=1}^{A} V(\boldsymbol{r}_i, \boldsymbol{r}_j) \,, \tag{3.1}$$

and tries to separate this into a Hamiltonian

$$H = \sum_{i=1}^{A} h_0(i) + H_{\text{residual}} \,, \tag{3.2}$$

in which $h_0(i)$ describes the single-particle Hamiltonian

$$h_0(i) = t_i + U(\boldsymbol{r}_i) \,. \tag{3.3}$$

The iterative (B)HF method, writing here only the direct term, then determines the single-particle field in terms of the nucleon–nucleon interaction. As an initial guess, one uses the convolution

$$U^0(\boldsymbol{r}_i) = \int \rho^0(\boldsymbol{r}_j) V(\boldsymbol{r}_i, \boldsymbol{r}_j) d\boldsymbol{r}_j \,, \tag{3.4}$$

where $\rho^0(\boldsymbol{r}_j)$ is the initial guess for the total nuclear density. The corresponding 'one-body' Schrödinger equation then becomes

$$-\frac{\hbar^2}{2m}\Delta\psi_a^{(1)}(\boldsymbol{r}_i) + U^0(\boldsymbol{r}_i)\psi_a^{(1)}(\boldsymbol{r}_i) = \varepsilon_a^{(0)}\psi_a^{(1)}(\boldsymbol{r}_i) \,, \tag{3.5}$$

for each single-particle state $\psi_a^{(1)}(\boldsymbol{r}_i)$ (with a denoting all the quantum numbers needed to describe the state uniquely). Having thus determined a first approximation to the wave function $\psi_a^{(1)}(\boldsymbol{r}_i)$, one can calculate an improved total density $\rho^{(1)}(\boldsymbol{r}_i)$ and then begin iterating until convergence in the energies $\varepsilon_a^{(n)}$, the wave functions $\psi_a^{(n)}(\boldsymbol{r}_i)$, and the potential $U^{(n)}(\boldsymbol{r}_i)$ results.

This procedure works well whenever the nucleon–nucleon interaction energy remains small compared to the nucleon rest mass. In the opposite case one has to resort to relativistic methods and we refer to Dirac–Hartree–Fock theories, Dirac phenomenology, etc., or, at the high-energy side, since one is probing further than just nucleonic degrees of freedom and is entering a new 'length scale', one will have to take both nucleon and meson degrees of freedom into account in an explicit way.

Hartree–Fock calculations have been carried out over the years for many mass regions. There exists a vast literature on this subject and, at the end of this chapter, a number of references will be given. Here, we show the resulting proton and neutron local potentials for the case of ^{208}Pb as obtained using the highly popular effective-density-dependent Skyrme nucleon–nucleon force (Fig. 3.3). This force contains, besides the regular two-body terms, also important contributions that are density dependent and thus effectively takes into account the many-body higher-order nucleon–nucleon interaction effects that inevitably appear within the nuclear medium. Furthermore, it has an additional dependence on the velocities of the interacting nucleons. This is a force which leads to saturation in the nuclear binding at the correct density, an important aspect that nuclear forces acting inside the atomic nucleus have to fulfil. If one were to use merely the nucleon–nucleon interaction that correctly describes free nucleon–nucleon scattering processes, the correct saturation properties would not be reproduced.

Starting now from the knowledge of these one-body nuclear properties (the mean field, the nucleon single-particle wave functions and energies) one can construct an approximate wave function describing, within this independent particle model description, the ground state of the atomic nucleus. This can be depicted as a fully antisymmetrized combination of the A individual wave functions and is denoted by

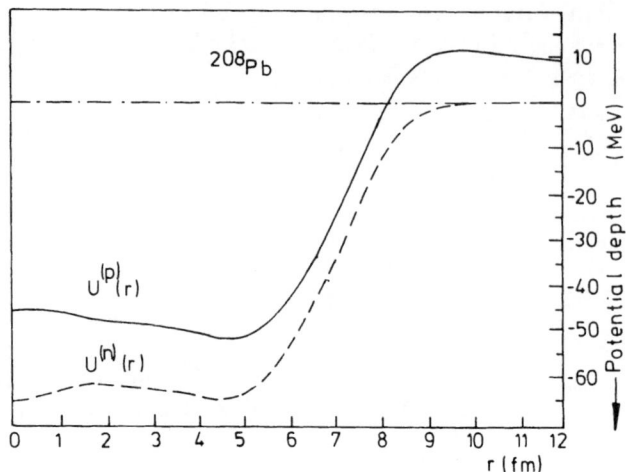

Fig. 3.3. The proton ($U^{(p)}(r)$) and neutron ($U^{(n)}(r)$) average potentials in a heavy nucleus (^{208}Pb), derived in a self-consistent way using a Skyrme effective density-dependent nucleon–nucleon interaction and Hartree–Fock methods. (Adapted from M. Waroquier et al. (1983) Nucl. Phys. A **404**, 298, Fig. 13. Elsevier Science, NL, with kind permission)

$$|\phi_{\mathrm{HF}}\rangle = \mathcal{A}\prod_{i=1}^{A}\psi_{a_i}(\boldsymbol{r}_i)|0\rangle \ , \tag{3.6}$$

with $\mathcal{A}$ the antisymmetrization operator.

Starting with this HF ground-state wave function, one can describe a number of properties such as the ground-state energy, or equivalently the nuclear binding energy

$$E_0 = \langle\phi_{\mathrm{HF}}|\hat{H}|\phi_{\mathrm{HF}}\rangle \ , \tag{3.7}$$

the nuclear mass (charge) distributions

$$\rho(\boldsymbol{r})_{(\pi,\nu)} = \langle\phi_{\mathrm{HF}}|\sum_{i=1}^{A}\hat{\rho}(\boldsymbol{r}_i)|\phi_{\mathrm{HF}}\rangle \ , \tag{3.8}$$

and nuclear radii (for mass and charge)

$$\langle r^2\rangle_{(\pi,\nu)} = \langle\phi_{\mathrm{HF}}|\sum_{i=1}^{A}r_i^2|\phi_{\mathrm{HF}}\rangle \ . \tag{3.9}$$

As a couple of examples we illustrate in Fig. 3.4 the nuclear charge densities for a number of nuclei ranging from the very light ^{4}He up to the heaviest doubly closed-shell nucleus ^{208}Pb. Not only are the distributions well described using present-day effective two-body forces; it is also clear that

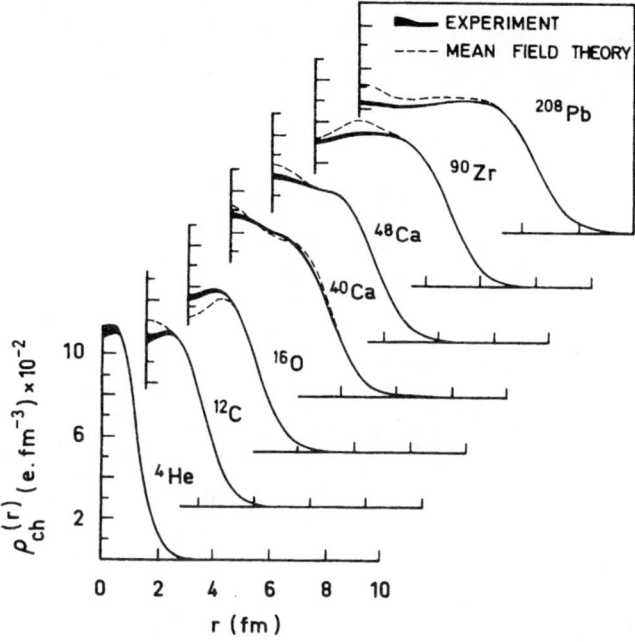

Fig. 3.4. Charge distributions for a number of doubly closed-shell nuclei. The thickness of the line indicates the experimental uncertainty in the data. The mean-field calculations were carried out by Decharge and Gogny (1968) Phys. Rev. **C21**, 1568. (Taken from Fig. 8, as given in B. Frois and C. Papanicolas (1987) Ann. Rev. Nucl. Part. Sci. **37**, 133)

saturation of the charge density indeed occurs. The central density barely varies when the nucleon number changes by more than an order of magnitude. What does change is the nuclear charge radius which also turns out to be a rather well-defined quantity. So the idea of regarding the nucleus as a charged, liquid drop when dealing with its static properties is well supported by the data. The same holds if we look at mass densities too (Fig. 3.5). Here, besides the mass densities for the range of nuclei between ^{16}O and ^{208}Pb we also indicate nuclear matter density (the latter assuming no surface effects are present or a surface contribution that is negligible compared to the volume effect). The figure for nuclear matter density has been obtained from HF calculations using Skyrme forces. In a separate technical box (Box I) we describe the basic features of this Skyrme force for those who like to see how this effective force actually looks.

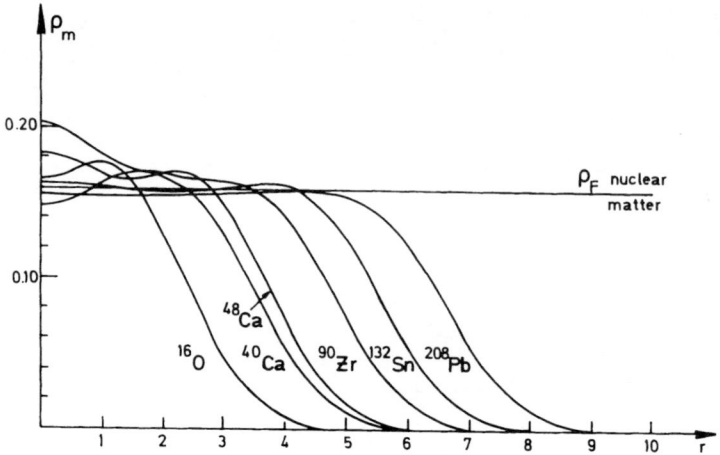

Fig. 3.5. Nuclear mass distribution for a number of nuclei with both a proton and neutron closed-shell configuration. The nuclear matter density is also given. The results are obtained using self-consistent calculations, starting from a Skyrme residual nucleon–nucleon interaction. (Taken from K. Heyde *The Nuclear Shell Model* ©1994 Springer, Berlin Heidelberg, with permission)

Box I

Extended Skyrme Forces (SkE) in Hartree–Fock Theory

In the extended Skyrme forces used, the two-body part contains an extra zero-range density-dependent term. In the three-body part, velocity-dependent terms are added. This is represented schematically:

Two-body part:
$$\nearrow V\left(\boldsymbol{r}_1, \boldsymbol{r}_2\right) = V^{(0)} + V^{(1)} + V^{(2)} + V^{(ls)} + V_{\text{Coul}} + (1 - x_3)V_0 \, ,$$

SkE

$$\searrow \text{Three-body part:}$$
$$W\left(\boldsymbol{r}_1, \boldsymbol{r}_2, \boldsymbol{r}_3\right) = x_3 W_0\left(\boldsymbol{r}_1, \boldsymbol{r}_2, \boldsymbol{r}_3\right) + W_1\left(\boldsymbol{r}_1, \boldsymbol{r}_2, \boldsymbol{r}_3, \boldsymbol{k}_1, \boldsymbol{k}_2, \boldsymbol{k}_3\right) \, .$$
$$(I.1)$$

Here, $V^{(0)}$, $V^{(1)}$, $V^{(2)}$ and $V^{(ls)}$ have the same structure as in the original Skyrme force parametrization [3.28], namely

$$V^{(0)} = t_0 \left(1 + x_0 P_\sigma\right) \delta\left(\boldsymbol{r}_1 - \boldsymbol{r}_2\right) \, ,$$
$$V^{(1)} = \frac{1}{2} t_1 \left[\delta\left(\boldsymbol{r}_1 - \boldsymbol{r}_2\right) \boldsymbol{k}^2 + \boldsymbol{k}'^2 \delta\left(\boldsymbol{r}_1 - \boldsymbol{r}_2\right)\right] \, ,$$
$$V^{(2)} = t_2 \boldsymbol{k}' \cdot \delta\left(\boldsymbol{r}_1 - \boldsymbol{r}_2\right) \boldsymbol{k} \, ,$$

$$V^{(ls)} = iW_0' \left(\boldsymbol{\sigma}_1 + \boldsymbol{\sigma}_2 \right) \cdot \left(\boldsymbol{k}' \times \delta \left(\boldsymbol{r}_1 - \boldsymbol{r}_2 \right) \boldsymbol{k} \right) , \tag{I.2}$$

where $\boldsymbol{k}$ denotes the momentum operator, acting to the right

$$\boldsymbol{k} = \frac{1}{2i} \left(\boldsymbol{\nabla}_1 - \boldsymbol{\nabla}_2 \right) , \tag{I.3}$$

and

$$\boldsymbol{k}' = -\frac{1}{2i} \left(\overleftarrow{\boldsymbol{\nabla}}_1 - \overleftarrow{\boldsymbol{\nabla}}_2 \right) , \tag{I.4}$$

acting to the left. The spin-exchange operator reads

$$P_\sigma \equiv \frac{1}{2} \left(1 + \boldsymbol{\sigma}_1 \cdot \boldsymbol{\sigma}_2 \right) , \tag{I.5}$$

and the Coulomb force has its standard form. The density-dependent zero-range force V_0 reads

$$V_0 = \frac{1}{6} t_3 \left(1 + P_\sigma \right) \rho \left(\left(\boldsymbol{r}_1 + \boldsymbol{r}_2 \right) / 2 \right) \delta \left(\boldsymbol{r}_1 - \boldsymbol{r}_2 \right) . \tag{I.6}$$

In the three-body part, one has the general term W_0

$$W_0 = t_3 \delta \left(\boldsymbol{r}_1 - \boldsymbol{r}_2 \right) \delta \left(\boldsymbol{r}_1 - \boldsymbol{r}_3 \right) , \tag{I.7}$$

to which a velocity-dependent zero-angle term W_1 is added,

$$\begin{aligned} W_1 = \frac{1}{6} t_4 \Big[&\left(\boldsymbol{k}_{12}'^2 + \boldsymbol{k}_{23}'^2 + \boldsymbol{k}_{31}'^2 \right) \delta \left(\boldsymbol{r}_1 - \boldsymbol{r}_2 \right) \delta \left(\boldsymbol{r}_1 - \boldsymbol{r}_3 \right) \\ &+ \delta \left(\boldsymbol{r}_1 - \boldsymbol{r}_2 \right) \delta \left(\boldsymbol{r}_1 - \boldsymbol{r}_3 \right) \left(\boldsymbol{k}_{12}^2 + \boldsymbol{k}_{23}^2 + \boldsymbol{k}_{31}^2 \right) \Big] . \end{aligned} \tag{I.8}$$

It can be shown that both interactions V_0 and W_0 contribute in the same way to the binding energy in even–even nuclei. The parameter x_3 has been retained so as to determine the pairing properties and thus, the properties of excited states near closed shells.

The spherical Hartree–Fock description used above hinges on numerical studies that can become rather involved. Moreover, once one goes away from the closed shell, binding energies need to be complemented by the extra energy gain due to nucleons interacting outside the closed shell systems (see also Sect. 3.2). Thus, a parametrization of the nuclear binding energy in the ground state described as a static liquid drop has been used and accurately describes the global behavior over vast ranges of nuclei (see the lower part of Fig. 3.6). This energy or mass formula, first discussed by Bethe, Bacher and Weizsäcker [3.33, 3.37], contains volume, surface, Coulomb, symmetry and a number of typical shell model correction terms (we neglect the last of these for the present purpose) and reads

$$B(E, A) = a_\text{v} A - a_\text{s} A^{2/3} - a_\text{c} Z(Z - 1) A^{-1/3} - a_A \frac{(A - 2Z)^2}{A} . \tag{3.10}$$

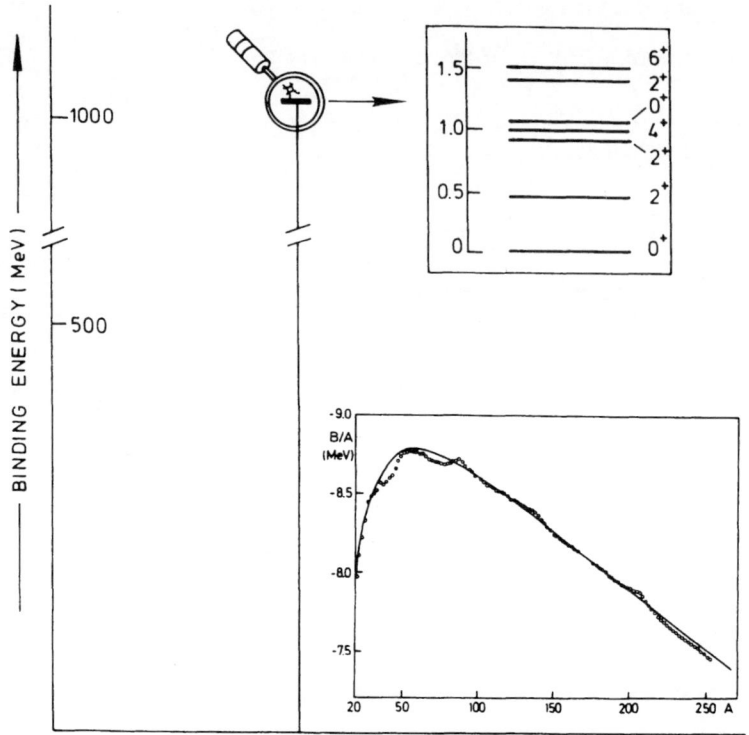

Fig. 3.6. Schematic illustration of a self-consistent calculation where, using a given nucleon–nucleon interaction, both the global nuclear characteristics (binding energy, radii, ...) as well as the local properties (detailed nuclear structure) (see *insert*) are derived in a consistent way. The difference in energy scale of three orders of magnitude is illustrated by using the magnifying glass. (Taken from K. Heyde *The Nuclear Shell Model* ©1994 Springer, Berlin Heidelberg, with permission)

By now extrapolating from a few hundred nucleons to systems containing only neutrons but adding also the gravitational binding energy term (neglecting surface effects and Coulomb contributions, but having a maximal symmetry energy effect) one arrives at a system for which the condition that it is still bound becomes

$$a_{\rm v} A - a_A A + \frac{3}{5} G \left(\frac{M^2}{r_0} A^{-1/3} \right) = 0 \, . \tag{3.11}$$

Solving for the number of nucleons needed to just fulfil the condition of obtaining a bound 'neutron nucleus', one obtains a value of $A \simeq 5 \times 10^{55}$ and, accordingly, a radius of a couple of kilometers. Objects of this form indeed exist as neutron stars. They can also be described in a deeper and more correct way: The neutron star is not just a neutron system: a more detailed look and a cut through a typical neutron star might look like the diagram shown in Fig. 3.7. It is important, however, to stress that it is mainly nuclear forces and the consequences of the strong forces that give rise to such stable objects which otherwise could not be explained easily.

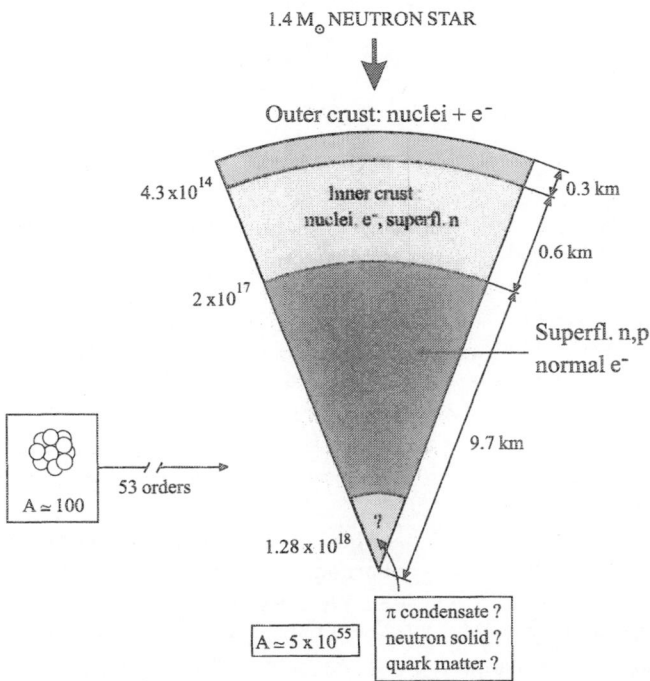

Fig. 3.7. Cut through a neutron star which can be viewed in a simplified way as a huge atomic nucleus containing only neutrons. The figures shows a more realistic distribution of matter (nuclei, electrons, neutrons, protons, other elementary particles, ...) for a neutron star of solar masses (1.4 $M_\odot$). (Taken from S. L. Shapiro and S. A. Teukolsky *Black Holes, White Dwarfs and Neutron Stars* ©1983 Wiley, N.Y., with permission)

Even though the above discussion has illustrated amply that correct saturation of the nuclear medium with good values for the nuclear binding energies and other bulk properties can be obtained using quite simple averaging methods like Hartree–Fock theory, one surely has to go beyond the 'global' scale and also study the 'local' fluctuations that give rise to the dynamical properties of the interacting many-body system. This is illustrated in Fig. 3.6 where the constraints one has to put on nucleon–nucleon effective forces inside the nucleus are such that:

(i) average properties must be described correctly, which means that for a nucleus with $A = 130$ nucleons, binding energies of the order of 1000 MeV will result, and,

(ii) detailed nuclear features which are three orders of magnitude smaller should at the same time be correctly described.

This very ambitious program sets stringent limits on the nucleon forces used. It can be shown that Skyrme forces can indeed bridge this large gap and, taken as an effective force, seem able to describe both long and short-range correlations in the nuclear potential energy structure.

3.2 The Nucleus as an Interacting Many-Body System

One of the major problems in gaining insight into how the many nucleons 'move' inside the nucleus is finding approximate solutions to the very complicated nuclear many-body problem whilst still keeping the basic physics clearly visible in simple concepts that one can easily comprehend.

The Hartree–Fock method is one such first-level approximation that carries a large number of nucleonic interaction processes over into a largely independent A-nucleon problem as described previously in Sect. 3.1. But even with a basic simple picture, and going from closed shells to closed shells, one cannot leave out the very important modifications that residual interactions bring into the original zeroth-order description.

An approximation that comes in at this point then attempts to transform the more complicated A-nucleon system governed by a basic nucleon–nucleon force into a much simpler "model" space where "model" interactions take over the role of the genuine nucleon–nucleon forces. This concept of mapping the original problem into a much simpler model problem, albeit with the constraints that the lowest energy eigenvalues map onto each other and that the model wavefunctions are projections of the more exact A-nucleon wavefunctions onto the model space, is a method used in other physical systems too: in atomic and molecular physics, solid state physics, and particle physics to name just a few of them. This mapping then also links the model effective forces to the original nucleon forces. We shall now outline the important steps of this process in more detail.

The method is very general in describing complex physical situations in a model concept and has been described in detail in a number of excellent papers (see the references [3.38]–[3.48] given at the end of this chapter). Asking that the lowest energy eigenvalues $E_i (i = 1, 2, \ldots \ldots)$ are identical in both the full (A-particle space) and the model space (called M-space with m as the dimension) and that wave functions are connected through a projection onto the model space, one can write the full and model wave functions as follows

$$\psi = \sum_{i=1}^{\infty} a_i \psi_i \,,$$

$$\psi^{M} = \sum_{i=1}^{m} a_i \psi_i \ . \tag{3.12}$$

The energy eigenvalue constraint now leads to an implicit equation for the model Hamiltionian and thus for the effective interaction acting in the model space since one can write

$$H\psi = E\psi \ ,$$
$$H^{M}\psi^{M} = E\psi^{M} \ . \tag{3.13}$$

In equating the full space to the model space expectation values one can solve for the model interaction in a perturbative way resulting in the expression

$$V^{M} = V + V\frac{\hat{Q}}{E - H_0} V^{M} \ , \tag{3.14}$$

in which V^{M} is the model interaction to be contrasted with the original two-body nucleon force V, and where $\hat{Q}$ projects out of the model space the intermediate states in the perturbation expansion. This mapping process is shown pictorially in Fig. 3.8. There is a price that one has to pay for going into this model space: The model force becomes dependent on the energy one is calculating and, generally, one obtains a non-Hermitian eigenvalue equation. These problems can be readily overcome with present-day techniques which lead towards tractable calculations, but at the expense of getting a full perturbation series for the model interaction which has to be truncated at some point. Thus one also needs to consider the convergence properties of this series. This has actually never been solved in a convincing way and the more operational method is to consider only the lowest-order corrections stemming from the smaller model space. We can do this by treating configurations where nucleons are lifted out of the filled orbitals (called hole (h) orbitals) into the valence space (called particle (p) orbitals). So, one includes the most simple two-particle two-hole (2p-2h) and one-particle two-hole (1p-2h) (or 1h-2p) configurations. In Fig. 3.9, we illustrate this process for the case of ^{18}O where just two neutrons move outside of the closed ^{16}O core in the regular sd shell model space. Here, the original two-body force that should be applied in conjunction with the full model space gives very bad agreement with the data when used in the small model space. Including the second-order corrections (called "particle-hole" bubble polarization corrections) a much improved agreement results.

Something one has to remember here is that one can start from a given two-body nucleon–nucleon force inside the nucleus to be used for the full space, but according to the actual model space chosen, one derives various model interactions, eventually arriving back at the original force. This concept of a model interaction is quite interesting since it facilitates an under-

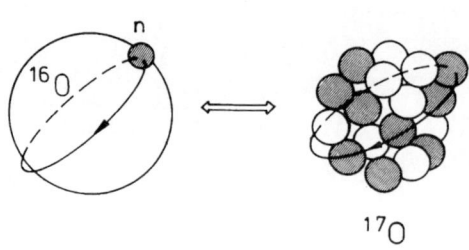

Fig. 3.8. Contrasting pictures of the nuclear many-body problem: *Left*, the model of a single particle, moving in the average field created by all other nucleons is shown; and *right*, the full complexity of the 17-particle problem. (Taken from K. Heyde *The Nuclear Shell Model* ©1994 Springer, Berlin Heidelberg, with permission)

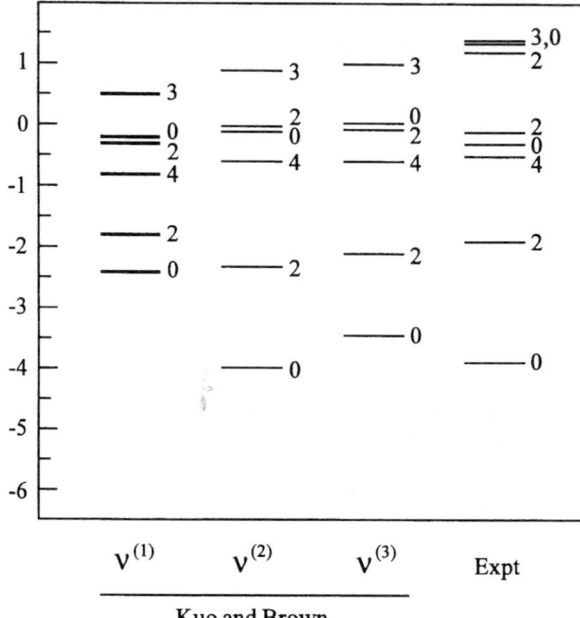

Fig. 3.9. Comparison of the experimental energy spectrum of the nucleus ^{18}O, with the results of different calculations using various model spaces. On the extreme left, purely two neutron interactions are considered within the sd-model space ($V^{(1)}$). In the other columns ($V^{(2)}, V^{(3)}$), corrections to this small model space constraint are added in second- and third-order perturbation theory. (Reprinted from M. Hjorth-Jensen et al. (1989) Phys. Lett. B **228**, 281. Elsevier Science, NL, with kind permission)

standing of salient nuclear structure effects where a large-scale calculation using the full space may not be so transparent.

The above method allowing us to determine model forces can also be used to obtain the model charges for the proton and the neutron as well as their model magnetic properties in a very similar way. Again asking that the corresponding matrix elements be mapped onto each other, as was done in determining the model force

$$\langle \psi^{\mathrm{M}} | \hat{O}^{\mathrm{M}} | \psi^{\mathrm{M}} \rangle = \langle \psi | \hat{O} | \psi \rangle \; , \tag{3.15}$$

this once more implies an implicit equation for the model operator, which can then be iterated through a perturbation theory expansion

$$\hat{O}^{\mathrm{M}} = \hat{O} + \hat{O} \frac{\hat{Q}}{E - H_0} V + \cdots \; . \tag{3.16}$$

Thus, model charges e_π, e_ν, $g_{s,l}(\pi, \nu)$ for proton and neutron, also quite often called effective charges are obtained. These act within the model space only and depend thereby on its dimension. In returning fully to the original large space one has to recover the bare charges and magnetic g-factors. This process can be visualized, as done in Fig. 3.10 for the magnetic properties. In a first step, quark degrees of freedom become modified (renormalized) into the free nucleon g-factors. Subsequently, for moments and other magnetic properties in actual nuclei, the particular model space used will imply a further modification of the free g-factors into model space g-factors. These can deviate quite strongly from the original bare values. As an example, for the electric quadrupole (E2) transition in ^{17}O in which, in the sd-model space, a neutron makes a transition from a $3\,s_{1/2}$ into a $2\,d_{5/2}$ single-particle state, an effective neutron electric charge of $e^{\mathrm{eff}} \simeq 0.4\,e$ is needed to comply with the observed lifetime of that particular $1/2^+$ level. The bare charges of $e_\pi = e$ and $e_\nu = 0\,e$ have to be considered when treating this same problem in the full complexity of the 17-particle problem in ^{17}O.

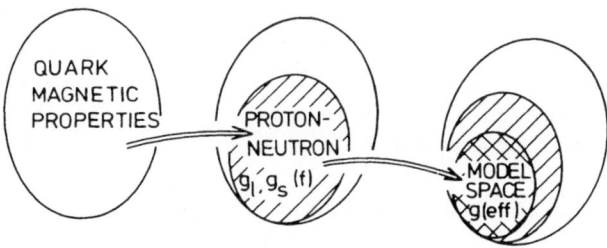

MAGNETIC PROPERTIES

Fig. 3.10. The successive mapping procedure for nuclear magnetic properties. At the *left*, we begin with the quark magnetic properties, proceeding then to effective nucleon (proton, neutron) single-particle g-factors, on into the full nuclear model space values for a given number of protons (Z) and nucleons (A)

Once the model properties are well determined, the remaining problem that must be solved in order to characterize the low-energy processes inside a given nucleus consists in solving the energy eigenvalue equation within the given model space. For few-particle problems (two or three nucleons outside a closed-shell system) or even when considering the simplest particle–hole excitations across a closed shell when describing closed-shell nuclei themselves (1p-1h), this is a simple problem. In each of the above cases, one constructs model configurations consistent with a given angular momentum and parity denoted as

$$\psi_{J\pi} = [\psi_{a_1}(\boldsymbol{r_1})\psi_{a_2}(\boldsymbol{r_2})]_{J\pi} \, |\phi_{\mathrm{HF}}\rangle \, , \qquad (3.17)$$

and

$$\psi_{J^\pi} = \left[\psi_{a_1}(r_1)\psi_{a_2}^{-1}(r_2)\right]_{J^\pi} |\phi_{HF}\rangle \,, \tag{3.18}$$

for a two-nucleon and one-particle one-hole system, respectively. Here, $|\phi_{HF}\rangle$ denotes the HF groundstate or model reference state. In both cases, also in more complex cases, the secular equation to be solved becomes

$$H\phi_{J^\pi}^i = E(i, J^\pi)\phi_{J^\pi}^i \,, \tag{3.19}$$

with

$$\phi_{J^\pi}^i = \sum_{k=1}^{m} a^i(k, J^\pi)\psi_{J^\pi}^k \,. \tag{3.20}$$

One thereby needs as input the model interaction and the various single-particle energies, and calculates relative to an inert core. Before presenting a few examples from recent very large-scale shell model studies, we recall the basic approximation and limitations:

(i) The full shell model space has been separated into an inert core and a model space that is expected to contain all of the ongoing processes that should describe the data. Configuratons that lie outside that model space are not just discarded: they are incorporated, in a perturbation-theoretical way, through the model interaction and the model or effective charges. When dramatic phenomena are observed, their original frequently lies outside the model space. Such phenomena are most often referred to as "intruder" processes; they cannot be handled in a perturbative treatment and so much care is needed to treat them consistently with the original shell model.

(ii) The shell model actually tests simple geometrical relations that determine how, starting from a given set of single-particle energies and two-body model matrix elements, these propagate into a many-body interacting system. The interaction energy roughly scales according to $n(n-1)/2$ for the two-body forces and according to n for the one-body energies. Thus small deviations between experiment and theory for few-particle systems very rapidly 'explode' when one proceeds to large particle numbers in mid-shell situations.

We illustrate some results for large-scale shell model studies in the sd shell (Brown and Wildenthal [3.49]) and for the fp shell (Otsuka [3.54]) in Figs. 3.11 and 3.12. For more extensive and detailed studies on how the model interactions have been determined by fitting to a large body of experimental data, see the references at the end of this chapter.

It should have become clear that for determining energy eigenvalues and the corresponding wave functions in order to calculate the many nuclear observables, one needs large and fast computing capacity. With the advent of large-scale advanced parallel computing at a number of computer installations, vast shell model studies have been undertaken in many mass regions, quite often with remarkable success. At the same time, one runs the risk of

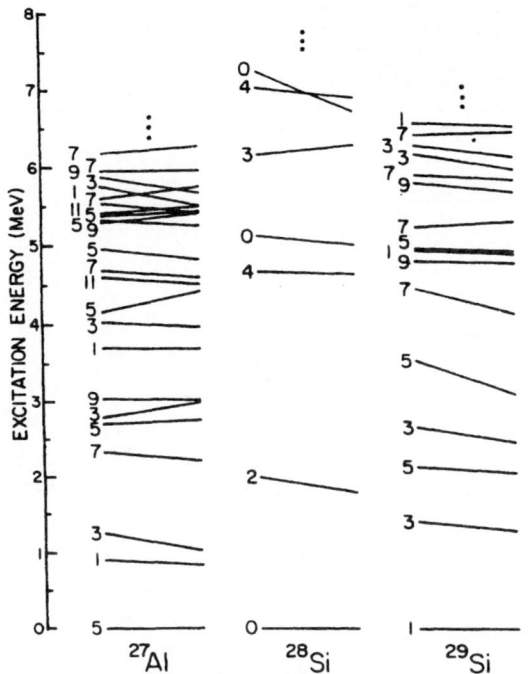

Fig. 3.11. Typical results for a large-scale *sd* shell model calculation for the ^{27}Al, ^{28}Si and ^{29}Si nuclei. The experimental (*right-hand side*) and theoretical (*left-hand side*) levels are joined by straight lines. (Reprinted from B. H. Wildenthal *Int. Symp. on Nuclear Shell Models* ed. by M. Valliéres, B. H. Wildenthal ©1985 World Scientific, Singapore, with permission)

losing the insight into which components of the nuclear force and of the model space actually explain why certain regular patterns of motion of the interacting protons and neutrons exist and at which particular particle numbers they appear. Clearly going to progressively higher energies inside the atomic nucleus, the distinction between a more-or-less inert core and a model space may progressively dissolve and the agreement achieved with the shell model deteriorate. Some of these effects have been observed: In the middle region of the *sd* shell where nuclei like ^{24}Mg, ^{28}Si, and ^{32}S appear, even though the energy spectra are quite well described, one cannot understand the electromagnetic properties using a single and constant set of model effective charges. They have clearly increased over their original model space valence value, signaling that important particle–hole excitations across the closed shell are modifying the original separable shell model problem.

A recent solution to this problem has come from a quite unexpected corner. Shell Model Monte Carlo (SMMC) methods have allowed 'sampling' of the larger, unconstrained full shell model spaces and give reason to hope that a number of aspects of the nuclear shell model problem can be solved. The essentials of this SMMC method are described succinctly in a technical box (Box II) and we refer the interested reader to some basic review articles on this subject.

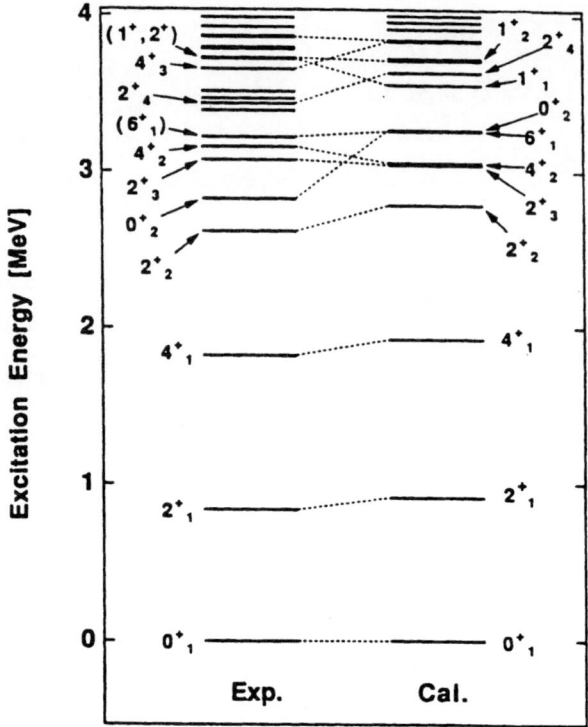

Fig. 3.12. The results from a large-scale fp shell model calculation carried out by Otsuka et al. A detailed comparison of theoretical results with the experimental data is made for ^{54}Cr. (Reprinted from H. Nakada et al. (1994) Nucl. Phys. A **571**, 467. Elsevier Science, NL, with kind permission)

Another way around these difficulties, which we shall discuss in more depth, takes a totally different look at the atomic nucleus as an interacting many-body fermion problem and concentrates more on the symmetries governing the interactions between nucleons and the implications for obtaining coordinated and collectively organized patterns. This will be the subject of Sect. 3.3.

Box II

The Shell Model Monte Carlo Method

The shell model Monte Carlo (SMMC) method is based on a statistical formulation. The canonical expectation value of an observable A at a given temperature T is given by the expression ($\beta = 1/T$)

$$\langle A \rangle = \frac{\text{Tr}_A \left(A e^{-\beta H} \right)}{\text{Tr}_A \left(e^{-\beta H} \right)} \ , \tag{II.1}$$

where $U = \exp(-\beta H)$ is the imaginary-time many-body propagator and $\text{Tr}_A U$ is the canonical partition function for A nucleons. The shell model Hamiltonian H can be cast in the form

$$H = \sum_\alpha \left(\epsilon_\alpha^* \bar{O}_\alpha + \epsilon_\alpha O_\alpha \right) + \frac{1}{2} \sum_\alpha V_\alpha \left\{ O_\alpha, \bar{O}_\alpha \right\} , \tag{II.2}$$

where ϵ_α are the single-particle energies and O_α represent a set of one-body density operators ($\bar{O}$ denotes the time-reverse of O). The Hamiltonian in (II.2) is manifestly time-reversal invariant if the parameters V_α that define the strength of the residual two-body interactions are real.

The key to the SMMC method is to rewrite the propagator U as a functional integral over one-body propagators. To achieve this goal, the exponent in U is split into N_t time slices of duration $\Delta\beta = \beta/N_t$,

$$U = \left[e^{-\Delta\beta H} \right]^{N_t} . \tag{II.3}$$

The many-body propagator at each time slice is linearized by a Hubbard–Stratonovich transformation, see [3.61], i.e., it is transformed into an integral over a set of one-body propagators that correspond to non-interacting nucleons in fluctuating auxiliary fields defined by complex c-numbers $\sigma_{\alpha n}$ ($n = 1,...,N_t$). The expectation value of A then reads

$$\langle A \rangle_A = \frac{\mathrm{Tr}_A \left(A e^{-\beta H} \right)}{\mathrm{Tr}_A \left(e^{-\beta H} \right)} \approx \frac{\int D[\sigma] W(\sigma) \Phi(\sigma) \langle A \rangle_\sigma}{\int D[\sigma] W(\sigma) \Phi(\sigma)} , \tag{II.4}$$

where the metric is

$$D[\sigma] = \prod_{\alpha n} \left[d\sigma_{\alpha n} d\sigma_{\alpha n}^* \Delta\beta |V_\alpha|/2\pi \right] , \tag{II.5}$$

and the approximation becomes exact as $N_t \to \infty$. The non-negative weight is

$$W(\sigma) = |\xi(\sigma)| \exp \left(-\frac{1}{2} \sum_{\alpha n} |V_\alpha| |\sigma_{\alpha n}|^2 \Delta\beta \right) , \tag{II.6}$$

where $\xi(\sigma) = \mathrm{Tr}_A U_\sigma$ is the partition function of the one-body propagator $U_\sigma = U_{N_t} \cdots U_1$, with $U_n = \exp(-\Delta\beta h_n)$, and the one-body Hamiltonian for the nth time slice is

$$h_n = \sum_\alpha \left(\epsilon_\alpha^* + s_\alpha V_\alpha \sigma_{\alpha n} \right) \bar{O}_\alpha + \left(\epsilon_\alpha + s_\alpha V_\alpha \sigma_{\alpha n}^* \right) O_\alpha , \tag{II.7}$$

with $s_\alpha = \pm 1$ for $V_\alpha < 0$ and $s_\alpha = \pm i$ for $V_\alpha > 0$. The "sign" is $\Psi(\sigma) = \xi(\sigma)/|\xi(\sigma)|$ and the expectation value of A with respect to the auxiliary field σ is

$$\langle A \rangle_\sigma = \mathrm{Tr}_A A U_\sigma / \xi(\sigma) . \tag{II.8}$$

Both $\xi(\sigma)$ and $\langle A \rangle_\sigma$ can be evaluated in terms of the matrix U_σ that represents the evolution operator U_σ in the space of N_s single-particle states. In applications, the trace is canonical corresponding to a nucleus with a fixed number of nucleons. Details of the transformation from the residual particle-particle interaction to the V_α used above can be found in the references on the SMMC method.

If all $V_\alpha < 0$, then the sign is $\langle \Psi \rangle = 1$. However, for realistic nuclear interactions such as Kuo–Brown, see [3.40], [3.45], [3.46], about half of the V_α's are positive, generating a sign problem (where the uncertainty in Ψ is larger than $\langle \Psi \rangle$). To overcome this problem, we extrapolate observables calculated for a family of good-sign Hamiltonians H_g (with $g < 0$) to the physical Hamiltonian at $g = 1$.

The SMMC calculations for fp-shell nuclei have been performed in the complete set of $1f_{7/2,5/2} - 2p_{3/2,1/2}$ configurations using the modified Kuo–Brown residual interaction. Each calculation involved 4000–5000 Monte Carlo samples at each of six values of the coupling constant g equally spaced between -1 and 0; extrapolation to the physical case ($g = 1$) was done by the method described by Koonin, Dean, and Langanke [3.61]).

3.3 Symmetries in the Atomic Nucleus

Because of the strong nuclear binding forces acting inside the whole atomic nucleus, the A localized nucleons that constitute the nucleus appear very much like a liquid drop that is able to undergo dynamical fluctuations around a spherical equilibrium shape. This then gives rise to structural symmetries (Sect. 3.3.1). The nuclear binding forces themselves also exhibit a number of symmetries that can subsequently give rise to symmetries within the interacting fermion system. This in turn may cause the appearance of dynamical symmetries under certain conditions (Sect. 3.3.2).

3.3.1 Structural Symmetries

The flow patterns in the motion of nucleons can exhibit collective coherent effects: ellipsoidal (quadrupole) small amplitude oscillations can be set up in the nuclear interior. Both in-phase and out-of-phase modes of motion can result and give rise to elementary excitations that optimally exploit the nucleon–nucleon interactions inside the atomic nucleus. These so-called structural symmetries relate to the fact that global flow patterns exist, which give rise to stationary modes (Fig. 3.13). A large number of these have been observed in the atomic nucleus and encompass both the low-lying quadrupole, octupole, and hexadecapole types of motion reaching into the much higher (10–20 MeV) region where giant resonances [resonances because the excitations at this energy are unstable against the emission of particles (proton, neutron, α-particle,...)] of various multipoles have also been detected. The giant electric dipole resonance (called GDR in abbreviated form) was one of the early ones to be detected but more complex modes like coherent spin (and isospin) modes where the spins and charges of the individual nucleons act cooperatively over the whole nucleus have been studied. More recently, identified in electron scattering experiments at Darmstadt and later explored using photon and hadron scattering, a 'scissors' mode in which protons and neutrons can carry out contra-rotational motion has been observed and mapped throughout the whole nuclear mass region (Fig. 3.14). References to these recent exciting developments are given at the end of this chapter, see [3.73]–[3.78].

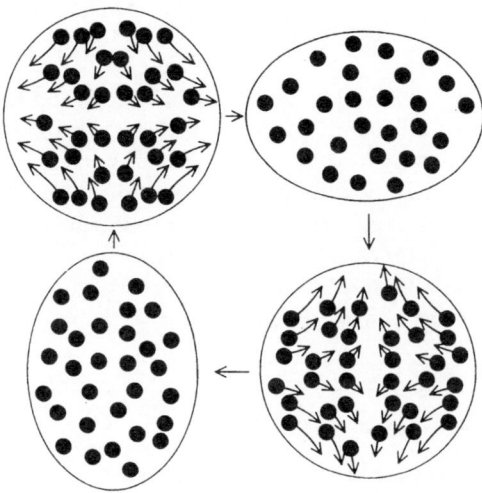

Fig. 3.13. Quadrupole vibrational excitations. The various phases of the oscillatory motion are depicted showing the flow patterns that redistribute the protons and neutrons, in phase, from a spherical into an ellipsoidal shape and back. (Adapted from G. Bertsch (1983) *Scientific American* May, by courtesy of J. Kuhl)

On the level of the nucleons themselves, invariance of the total nuclear wave functions under the exchange of identical nucleons affects the possible modes realized: the Pauli principle implies antisymmetry for the total fermionic wave function and this has very definite consequences for the types of collective motion that can be set up inside the nucleus.

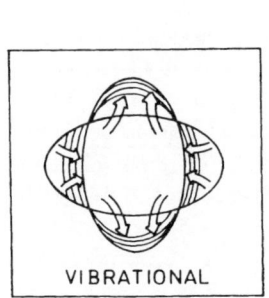

VIBRATIONAL

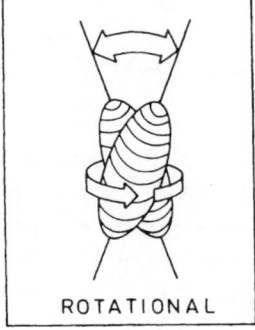

ROTATIONAL

Fig. 3.14. Schematic drawing of elementary modes of motion in which protons and neutrons are moving in a non-symmetric way. The two examples illustrate a non-symmetric vibrational and a non-symmetric rotational 'scissors' type of motion. (Taken from K. Heyde *Basic Concepts in Nuclear Physic* ©1994 IOP Publishing, with permission)

3.3.2 Dynamical Symmetries

The nuclear two-body force obeys quite a large number of basic invariances (invariance under interchange of the spatial coordinates, translation invariance, Galilean invariance, space reflection symmetry, time reversal invariance, rotational invariance in coordinate space, rotational invariance in charge space, or isospin symmetry, ...)(Chap. 2). These invariances or symmetries give rise to conserved quantities, as is well known from basic quantum mechanics.

Here, we would like to point out that the symmetries of the interacting system, described by a particular Hamiltonian, have been used throughout to classify and keep order in the variety of nuclear excitations modes. In a technical box on symmetries (Box III), we illustrate a number of the key symmetries that have played a dominant role, starting from the isotopic symmetry [SU(2) symmetry] and Wigner supermultiplet symmetry [SU(4)] up to recent symmetries related to various collective modes of motion encompassing the many nucleons that are present inside the atomic nucleus.

Box III

Symmetry Concepts in Nuclear Physics

Here we consider briefly how the concept of symmetries in physics has always been a guideline in unifying seemingly unconnected phenomena. Let us simply list some of the major advances made using symmetry concepts in describing different aspects of the nuclear many-body system.

1932: The concept of isospin symmetry, describing the charge independence of the nuclear forces by means of the isospin concept with the SU(2) group as the underlying mathematical group was suggested by Heisenberg [3.79]. This is the simplest of all dynamical symmetries and expresses the invariance of the Hamiltonian under the exchange of all proton and neutron coordinates.

1936: Spin and isospin were combined by Wigner [3.80] into the SU(4) supermultiplet scheme with SU(4) as the group structure. This concept has been extensively used in the description of light α-like nuclei ($A = 4 \times n$).

1942: The residual interaction amongst identical nucleons is particularly strong in $J^{\pi} = 0^+$ and 2^+ coupled pair states. This "pairing" property is a cornerstone in accounting for the nuclear structure of many spherical nuclei, particularly those near closed shells. Pairing is at the origin of seniority, itself related to the quasi-spin classification and group, as used first by Racah [3.81] in describing the properties of many-electron configurations in atomic physics.

1948: The spherical symmetry of the nuclear mean field and the realization of its major importance for describing the nucleon motion in the nucleus was put forward by Mayer, Haxel, Jensen, and Suess [3.82], [3.83].

1952: The nuclear deformed field is a typical example of the concept of spontaneous symmetry breaking. The restoration of the rotational symmetry, present in the Hamiltonian, leads to the formation of nuclear rotating spectra. These properties were discussed earlier in a more phenomenological way by Bohr and Mottelson [3.84]–[3.86].

1958: Elliott [3.87], [3.88] remarked that in some cases, the average nuclear potential could be depicted by a deformed, harmonic oscillator containing the SU(3) dynamical symmetry. This work opened the first possible connection between the macroscopic collective motion and its microscopic description.

1974: Dynamical symmetries were introduced by Arima and Iachello [3.89], [3.90] in order to describe nuclear collective motion starting from a many-boson system with only s ($L = 0$) and d ($L = 2$) bosons. The relation to the nuclear shell model and its underlying shell structure has been studied extensively. These boson models have given new momentum to nuclear physics research.

The above symmetries are depicted schematically in Fig. III.1.

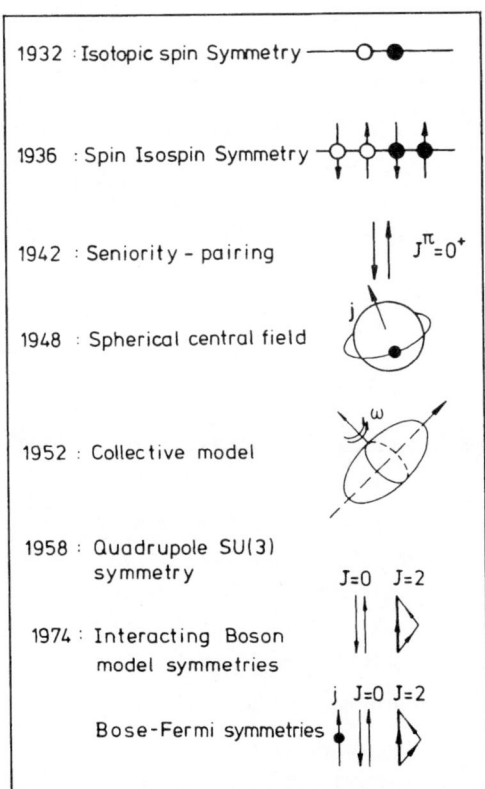

Fig. III.1. Symmetries in nuclear physics. (Taken from K. Heyde *The Nuclear Shell Model* ©1994 Springer, Berlin Heidelberg, with permission)

Let us consider the simple example of integer angular momentum characterized by ℓ and its magnetic quantum number m. These two quantum numbers are related to the group chain

$$O(3) \supset O(2)$$

$$\vdots \quad \vdots$$

$$\ell \quad m$$

and the operators $\hat{\ell}^2$ and $\hat{\ell}_z$ are the invariant or Casimir operators for the corresponding O(3) and O(2) groups. Now, any Hamiltonian that can be written using the invariant operators $\hat{\ell}^2$ and $\hat{\ell}_z$ only can be diagonalized within the basis states that characterize the O(3) $\supset$ O(2) group reduction. As an example we consider the Hamiltonian

$$\hat{H} = \alpha \left[\hat{\ell}^2 - \hat{\ell}_z \left(\hat{\ell}_z - 1 \right) \right] . \tag{3.21}$$

Such a Hamiltonian is said to exhibit a 'dynamical symmetry'. The various multiplet members for a given set of quantum numbers (ℓ, m) will then be split by the Hamiltonian but no mixing of the quantum numbers results. The eigenvalue for the Hamiltonian of (3.21) then becomes

$$E(\ell, m) = \alpha \left[\ell \left(\ell + 1 \right) - m \left(m - 1 \right) \right] . \tag{3.22}$$

This same idea can now be generalized to a much larger group G and the problem of studing the dynamical symmetries can be posed in terms of this group. One has to:

(i) construct a basis by looking for all subgroups G' of the largest group G;
(ii) analyze the quantum numbers that are associated with the various subgroups G';
(iii) construct and study the eigenvalues of the various invariant (Casimir) operators for the groups G';
(iv) diagonalize a given Hamiltonian.

For a Hamiltonian that is written in terms of the invariant operators of a given group chain only

$$H = \alpha C(G) + \alpha' C(G') + \alpha'' C(G'') + \cdots , \tag{3.23}$$

then the energy is obtained as

$$E = \alpha \langle C(G) \rangle + \alpha' \langle C(G') \rangle + \alpha'' \langle C(G'') \rangle + \cdots , \tag{3.24}$$

and it is said that the quantum system exhibits a dynamical symmetry.

To illustrate the idea of dynamical symmetries, Figs. 3.15–3.17 show:

(i) the excitation modes of a spherical harmonic oscillator shell (with $N = 4$) where both orbital angular momentum and spin–orbit terms have been added (Fig. 3.15, left-hand part);

(ii)   the example of Zeeman splitting of spherical shell model states in an external magnetic field, depending on $\hat{j}_z$ only (Fig. 3.15, right-hand part);

(iii) the energy spectrum obtained from the general two-body Hamiltonian describing an interacting system of Z protons and N neutrons where both the small mass differences between protons and neutrons as well as the Coulomb forces are included (Fig. 3.16);

(iv)   the mass spectrum of the SU(3) octuplet of particles including the proton and neutron states (Fig. 3.17).

Many more examples can be found in the specialized literature.

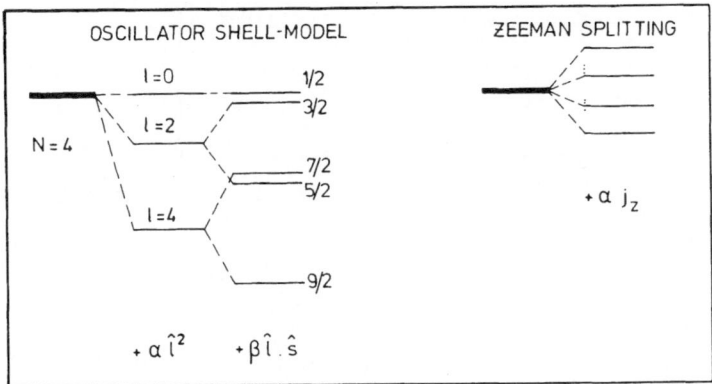

Fig. 3.15. The dynamical symmetry for the harmonic oscillator model ($N = 4$) for which the degeneracies are broken in a systematic way by adding the orbital $\hat{l}^2$ and the spin–orbit $\hat{l}.\hat{s}$ terms. The quantum numbers $N(l, 1/2)j$ remain good quantum numbers. On the *right-hand side*, the addition of a Zeeman term $(\hat{j}_z)$ then splits the remaining m-degeneracies of the spherical single-particle field. (Taken from K. Heyde *The Nuclear Shell Model* ©1994 Springer, Berlin Heidelberg, with permission)

A very interesting example of the application of these powerful group-theoretical methods in the study of low-lying nuclear quadrupole motion the approximation known as the interacting boson model (IBM), in which interacting pairs of nucleons coupled to angular momentum 0^+ and 2^+ only (the most strongly bound nucleonic pair states inside the nucleus) are considered and are treated as genuine bosons (Fig. 3.18). This interacting system of s and d bosons forms a U(6) group-theoretical structure possessing various possible reduction schemes. These have been studied in many technical detail in a number of review articles, monographs, and books given in the reference list at the end of the this chapter. Here we illustrate (Fig. 3.19) the three basic reduction chains reducing U(6), into the O(3) subgroup via the U(5), O(6), or SU(3) subgroup chains, giving rise to quite important differences in the way the interacting N-boson system is organized. In Fig. 3.20, we show the example of the U(5) reduction which corresponds to typical anharmonic quadrupole vibrational modes when one seeks a geometric analogy.

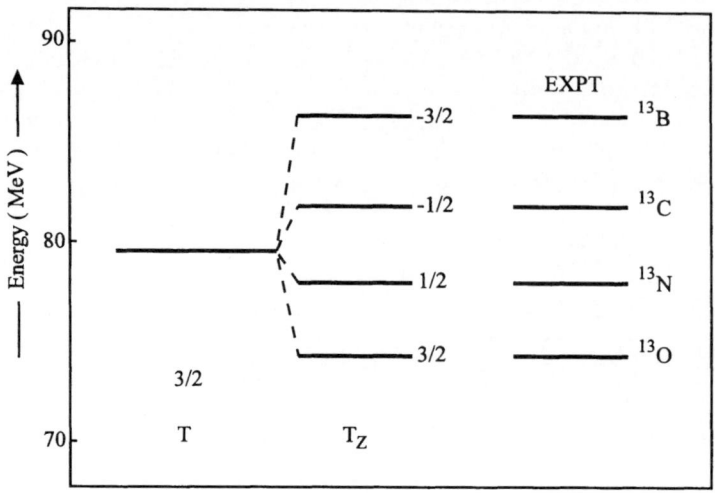

Fig. 3.16. Illustration of the dynamical symmetry SU(2), describing the interaction of both protons (including Coulomb forces) and neutrons, for the $A = 13$ quadruplet of states with isospin $T=3/2$ and projection $-3/2 \leq T_z \leq 3/2$. The energy equation $E(T,T_z) = a(T) + b(T)T_z + c(T)T_z^2$, describes the eigenvalues $[a(T) = 80.59\text{MeV} , b(T) = -2.96\text{MeV} , c(T) = -0.26\text{MeV}]$. (Taken from P. Van Isacker et al. (1994) J. Phys. G. **20**, 853. IOP Publishing, with permission)

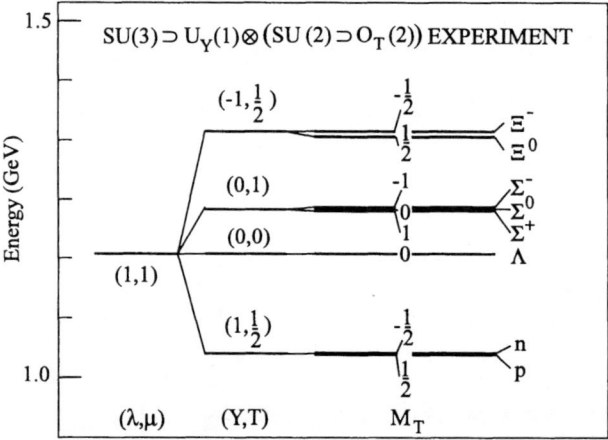

Fig. 3.17. The mass spectrum for the octuplet of particles (p,n), (Λ), $(\Sigma^0, \Sigma^\pm)$, (Ξ^{-0}) described as representations of the SU(3) dynamical symmetry. The degeneracy breaking is described by the SU(3) $\supset$ U$_Y$(1)$\otimes$(SU$_T$(2) $\supset$ O$_T$(2)) group chain, and the corresponding mass equation is $a+bY+d(T(T+1)-\frac{1}{4}Y^2)+eM_T+fM_T^2$ ($a = 1111.3$ MeV, $b = -189.6$ MeV, $d = -39.9$ MeV, $e = -3.8$ MeV, $f = 0.9$ MeV) (Taken from P. Van Isacker et al. (1994) J. Phys. G. **20**, 853. IOP Publishing, with permission)

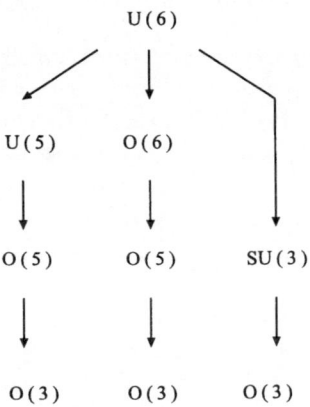

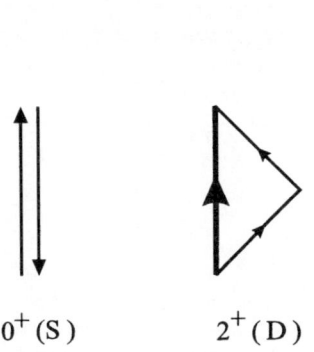

Fig. 3.18. The favored nucleon pair configuration (0^+ pair or S-pair) and the 2^+ or D-pair that form the underlying structure of the dynamical U(6) symmetry in nuclear physics

Fig. 3.19. The three possible reduction schemes from U(6), passing through the U(5), O(6) and SU(3) subgroup chains giving rise to the three distinct dynamical symmetry limits of the interacting boson model algebraic structure

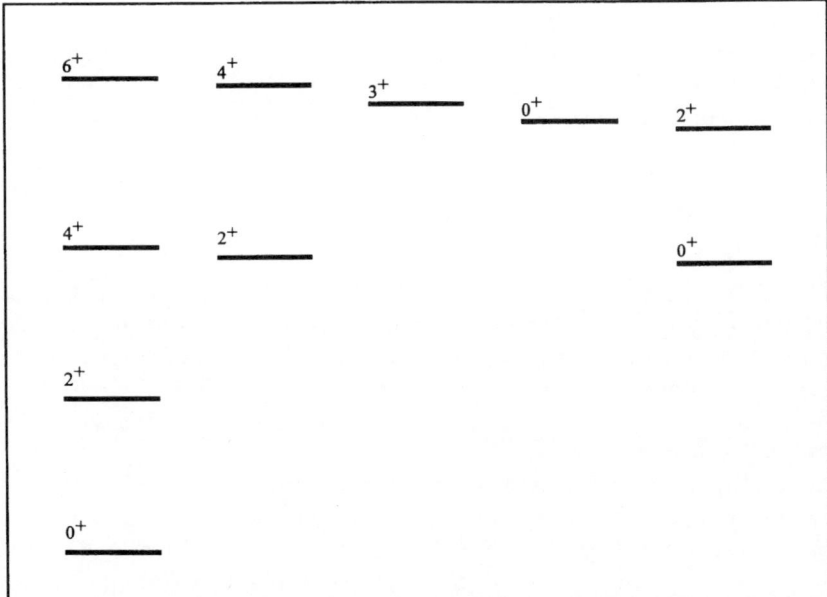

Fig. 3.20. Illustration of the U(5) or anharmonic quadrupole vibrational limit, which describes many nuclei with only a few nucleons moving outside a closed-shell configuration

The energy expression is particularly simple in this limit: it is characterized by only four parameters but nonetheless has a very large region of applicability. The other two limits, the SU(3) and O(6), can be related to axial rotational motion and gamma-soft rotor spectra, respectively. We refer to the extensive literature on this subject for further details at the end of this chapter.

3.3.3 Conclusion

The field of symmetries, be it just the structural symmetries that can be associated with certain flow patterns, or the symmetries that are more directly connected to the dynamics of the nucleon–nucleon interactions reflected in the Hamiltonian of the nuclear system, presents a very rich source of information. It has been and still is used as a major guiding principle for comprehending and bringing order into the interacting nuclear many-body system.

3.4 The Nuclear Structure Phase Diagram

3.4.1 Introduction

The atomic nucleus with its A nucleons is governed by a large number of degrees of freedom. From this multitude one can single out a rather small number of variables that determine the major characteristics and can also be regarded as external 'parameters', which, when altered, cause the atomic nucleus to respond in a specific way. These variables can then be used to characterize some major 'axes' defining a space in which one can explore and map the nuclear behavior as a multidimensional system. Moreover, these major 'axes' can be directly connected to recent research efforts and to the actual experimental possibilities for probing and detecting nuclear phenomena:

(i) As a first axis we can take the nuclear angular momentum. This can be influenced in heavy-ion reactions in which an ion is accelerated and impinges at grazing conditions on a target nucleus, thereby transferring a large amount of angular momentum. This can modify the motion of individual nucleons in an important way since they will start moving in a rotating deformed field.

(ii) A second axis can be related to the internal temperature and to heating the nuclear many-body system. Heating may cause the internal occupation of the various orbitals to be strongly modified and may give rise to dramatic changes in the internal structural 'organization' of nucleons inside the nucleus.

(iii) For a third axis, one moves outside of the region of beta stability and explores the edges of stability in the nuclear landscape. Here, a typical variable might be the ratio $(N - Z)/A$ (relative neutron excess). This

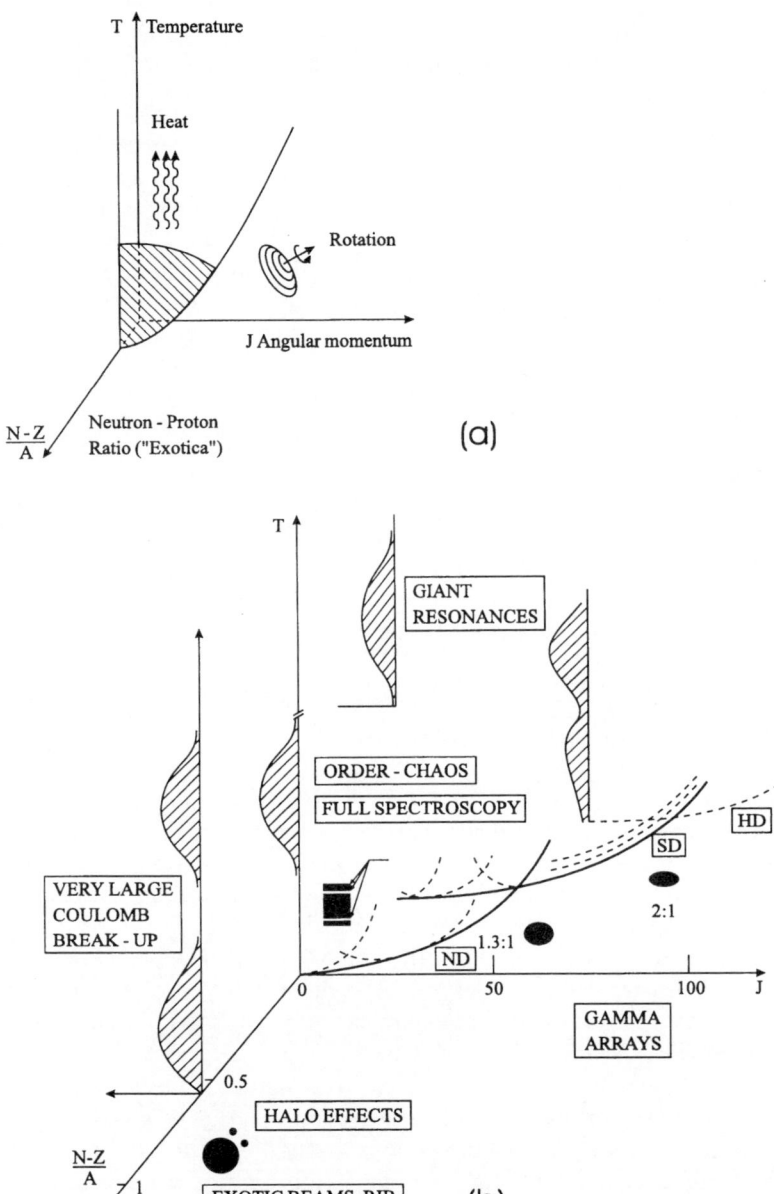

Fig. 3.21. (a) Schematic representation of nuclear structure in the space of three major axes describing some of the most important variables characterizing nuclear properties. The axes are the nuclear temperature axis, T, the nuclear rotational degree of freedom, J, and the neutron-to-proton ratio described via the variable $(N-Z)/A$ (relative neutron excess). **(b)** A more detailed version of the $(T, J, (N-Z)/A)$ axes partitioning nuclear structure properties. A variety of nuclear structure effects and the methods used to explore these three main directions are indicated. (Adapted from A. Richter (1993) Nucl. Phys. A **533**, 417c, Fig. 2. Elsevier Science, NL, with kind permission)

exploration is at present at one of the frontiers of research in forming and accelerating radioactive ion beams and a number of facilities are currently under construction or in planning (see also Chap. 7).

Of course, such a division with three major axes is not unique and not all phenomena can be described unambiguously in such a phase diagram. But these three axes are good starting points to explore the nucleus and we shall now look at three different trajectories in this "space". In Fig. 3.21 we show both a very schematic division (a) and a more detailed version (b) where a large number of nuclear phenomena currently being studied can be positioned.

3.4.2 Behavior of Rapidly Rotating Nuclei

Almost spherical nuclei in the vicinity of closed shells in which nucleons are preferentially found in 0^+ coupled pairs show high rigidity against deformation of any type. Before studying the properties of the nucleus when rapid rotation is imposed on its constituent nucleons, we shall bring in a technical box (Box IV) that explains the basic issues that describe a nucleon moving in a deformed field that is subsequently set into rotation.

Box IV ==

Nuclear Deformation and Rapid Rotation

When a nucleon is moving in an axially deformed quadrupole average field (e.g., a three-dimensional harmonic oscillator potential with two equal frequencies in the x, y plane but a different one in the z-direction), the otherwise $2j + 1$-fold degeneracy is split into $j + \frac{1}{2}$ two-fold degenerate orbitals (Fig. IV.1). Here only the projection of the single-particle angular momentum remains a good quantum number. In evaluating now the total energy of the nucleus as a function of the shape of the deformed potential, one may think of adding up all individual single-particle energies corresponding to the occupied states. This is not fully correct and a much used method relies on renormalizing the bulk energy to the liquid drop model value for the same deformation but now adding the local fluctuating energy term that arises from the non-uniform level distribution near to the Fermi level (Fig. IV.2).

This latter term is also called the shell-correction energy and when added to the global liquid drop term gives a good prescription for determining the total nuclear energy

$$E(\epsilon_2) = E_{\text{Liq.drop}}(\epsilon_2) + \delta E(\epsilon_2) \,. \tag{IV.1}$$

It allows one to determine equilibrium deformed shapes for many nuclei in regions where large quadrupole deformations appear: rare-earth nuclei, actinides, etc.

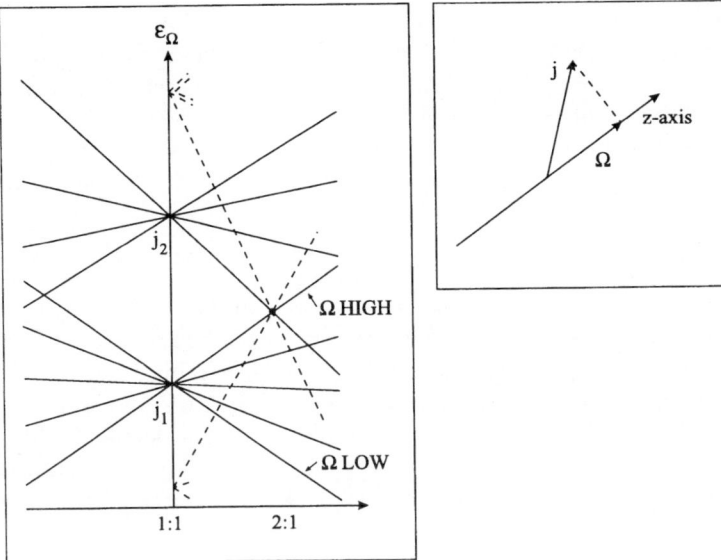

Fig. IV.1. Illustration of the energy levels, corresponding to an anisotropic harmonic oscillator potential, for which the $2j + 1$-fold degeneracy of the spherical field is split into $j + 1/2$ two-fold degenerate substates. On the abscissa 1 : 1 and 2 : 1 mean the ratio of major to minor axis. Here, Ω denotes the magnetic quantum number; the projection of j onto the z symmetry axis is illustrated *on the right*

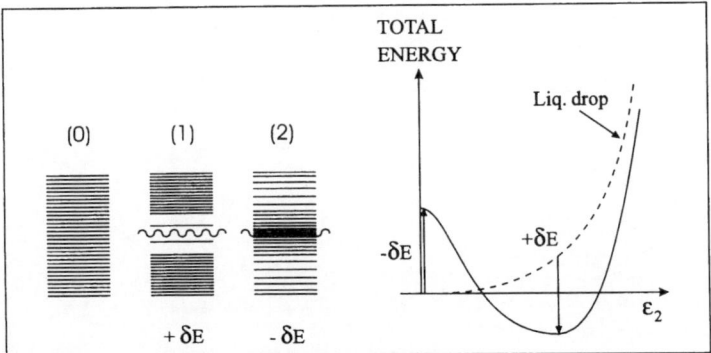

Fig. IV.2. *Left*: Various possible distributions corresponding to a nucleon moving inside the atomic nucleus in a deformed mean field. In the case 1 (2), increased (decreased) stability compared to the reference, regular level distribution (0) results at the position of the Fermi level. *Right*: This energy correction (denoted by δE) modulates the liquid drop model total energy value and may well result in stable, deformed minima in the total energy surface

One can now add the external rotational term which will break the time-reversal invariance that was present in the non-rotating field. Depending on the orientation of the angular momenta in the various substates relative to the external rotation field (which is put along the x-axis for convenience), this splitting will also affect the total nuclear energy, which now becomes not only a function of deformation but also of rotational frequency (Fig. IV.3). Using the same Strutinsky method, the resulting expression is

$$E(\epsilon_2, \omega) = E_{\text{Liq.drop}}(\epsilon_2, \omega) + \delta E(\epsilon_2, \omega) , \tag{IV.2}$$

which can be minimized in deformation space. One can thus follow the trajectory of a given nucleus in its minimum energy state as a function of both the deforming and rotating external agents. A very interesting result is that in, e.g., the rare-earth nuclei, minima have been obtained at high spin and near to an axis ratio of 2 : 1 (major to minor axis), called superdeformed states. Ample experimental evidence for such excitations has accumulated during recent years and some excellent review articles have been written on the subject [3.115]–[3.123].

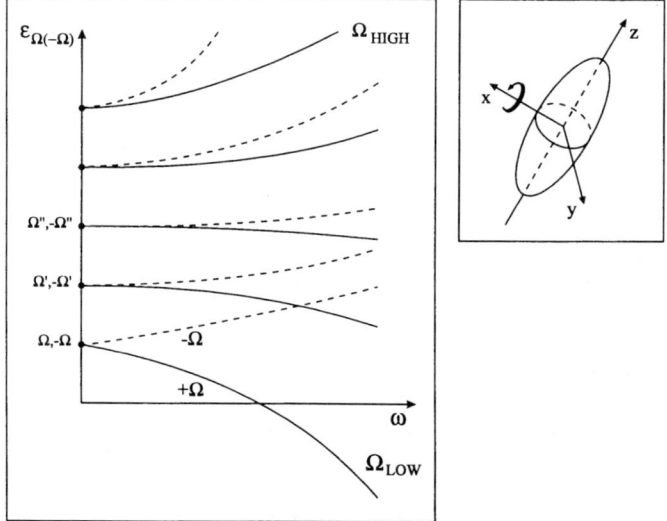

Fig. IV.3. Modification in the two-fold degeneracy $(\Omega, -\Omega)$ when an external potential is added by rotating the atomic nucleus around the x-axis (perpendicular to the symmetry z-axis of the nucleus) with a frequency ω. This extra term $\omega \hat{j}_x$ lifts the Kramers' (two-fold) time-reversal degeneracy in the single-particle motion

As was discussed in Box IV, nuclei can acquire a superdeformed energy minimum at high rotational frequencies. We first give some examples of how the total energy evolves as a function of quadrupole deformation and as a function of angular momentum of the nucleus for ^{132}Ce and ^{152}Dy (Fig. 3.22). A more elaborate theoretical study of ^{152}Dy is shown in Fig. 3.23 as a three-dimensional energy surface where, besides quadrupole deformation, the hexadecapole deformation degree of freedom is also taken into account. Figure 3.23 is constructed at angular momentum spin 80$\hbar$ and exhibits the state-of-the art in producing such landscapes.

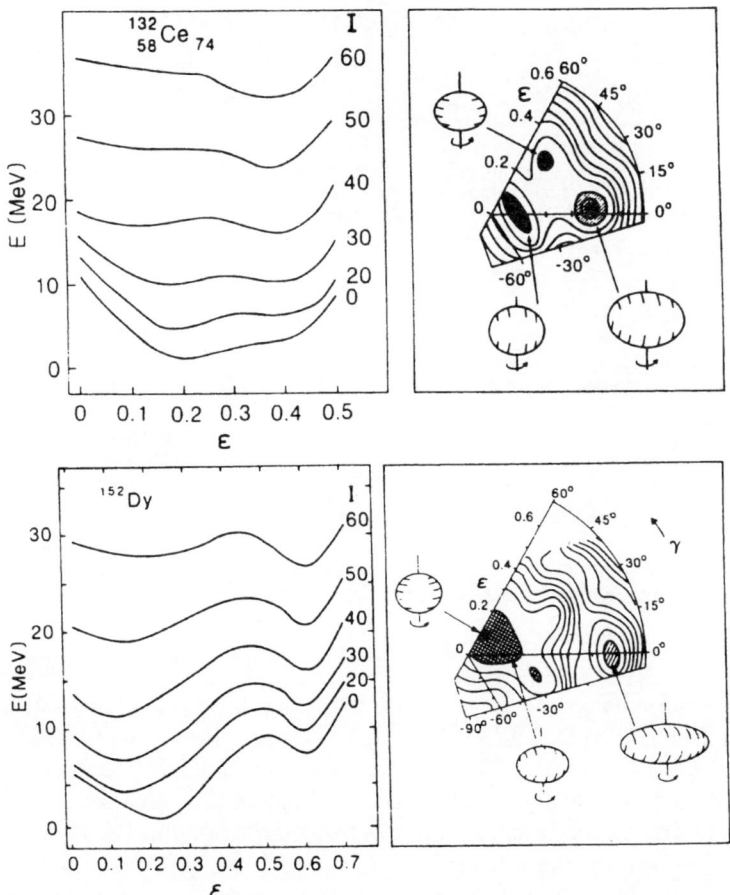

Fig. 3.22. Potential energy curves as a function of both the quadrupole deformation (ellipsoidal deformation and shape), characterized by the variable ϵ_2, and the angular momentum, characterized by the spin value J. On the *left-hand side*, the results are given for the deformed nuclei ^{132}Ce and ^{152}Dy. On the *right-hand side*, the various energy minima (and corresponding deformed shapes) are shown at a fixed angular momentum of spin 40$\hbar$ in the same two nuclei. (Taken from *The Annual Review of Nuclear & Particle Science* ©1988 Vol. 38, Annual Reviews Inc.)

Total energy: $^{152}_{66}Dy_{86}$

$I^{\pi} = 80^+$ $\beta_4 = minim.$

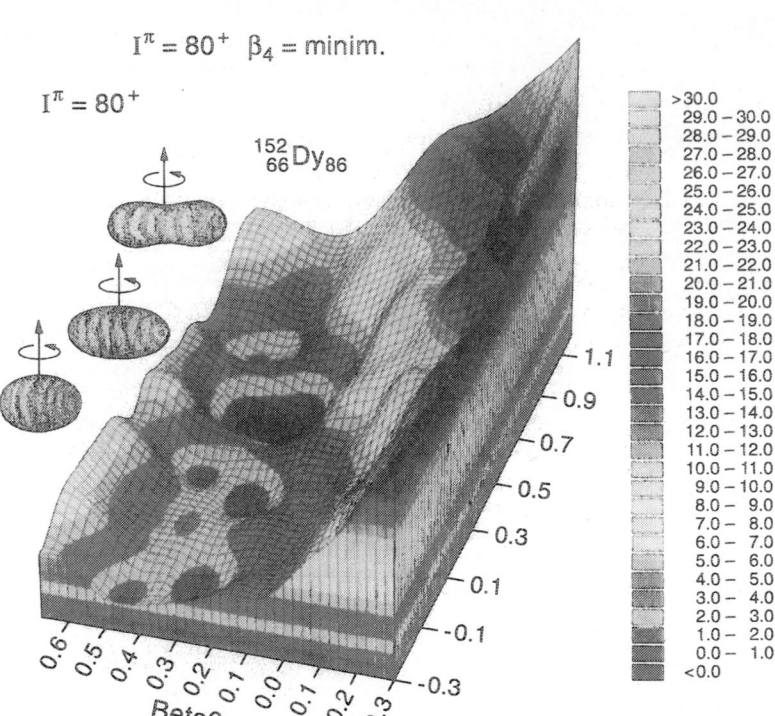

Fig. 3.23. Three-dimensional total energy surface for ^{152}Dy (see also Fig. 3.22) at spin $80\hbar$. Besides quadrupole deformation, hexadecapole deformation also is allowed to occur. The color code indicates the total energy scale. The specific deformed, rotating shapes at the three distinct minima are also shown. (By courtesy of J. Dudek)

One may wonder how such large amounts of angular momentum can be given to a nucleus. A typical method is 'fusion-evaporation' which consists in accelerating a medium-heavy fragment into a target nucleus (we illustrate this, in Fig. 3.24, for the case of ^{40}Ar accelerated into ^{124}Sn) at grazing angles and with high velocity such that the resulting compound nucleus (in this case ^{164}Er) is set into rapid rotation. This system is in general "hot" and will evaporate a number of nucleons (in our example four neutrons) before ending up as a bound nucleus ^{160}Er, which then cools further through the emisson of gamma radiation. In the course of this cooling process, the emitted gamma rays contain information about the cooling route and the physics of the region the nucleus is passing through. In order to find out about this decay route one should detect all the gamma rays as efficiently as possible using gamma-spectroscopic tools. A special 4π geometry has been set up

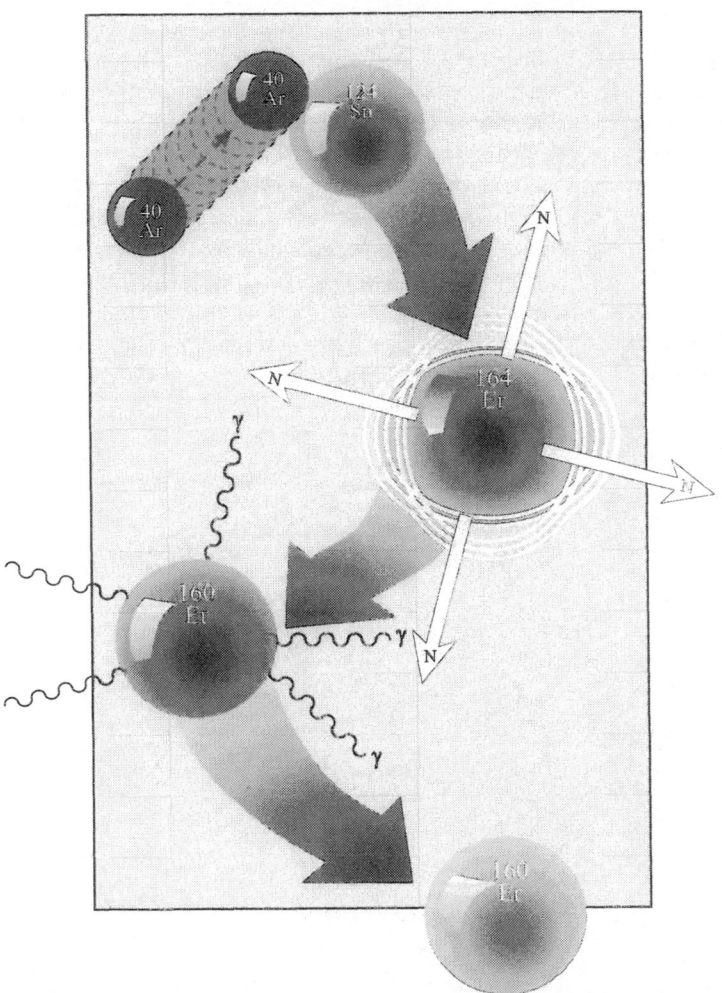

Fig. 3.24. The various steps in the reaction ^{124}Sn(^{40}Ar,3nγ)^{160}Er: (i) the ^{40}Ar + ^{124}Sn initial reaction, (ii) the fusion of the two nuclei into a compound system ^{164}Er, (iii) the emission of four neutrons from the rapidly spinning compound system, and (iv) the final cooling of ^{160}Er via gamma-ray emission. (Adapted from J. Goldhaber (1991) LBL Research Review, Vol. 16, No. 1, p. 24, with kind permission)

in order to reach a very high efficiency in detecting information that will allow a reconstruction of the decay path. A big "ball" of highly efficient Ge detectors with BGO shielding is set up around the reaction point. A number of such systems are now active, e.g., Gammasphere (Fig. 3.25), which has been operating at Lawrence Berkeley Laboratory for quite some time and is now moving to Argonne National Laboratory. These detectors (containing up to 110 elementary units) are very expensive and used mainly by relatively large (still rather small compared to a typical-sized high-energy experiment at one of the CERN LEP sites) collaborations amongst a number of laboratories. Thus, the ball is set up in such a way that it can be transported from one accelerator to another with the aim of studying different mass regions.

The first discovery of superdeformed bands in the decay of very rapidly rotating nuclei was made at the Daresbury Nuclear Structure Facility (which is now closed!!!) by a team headed by P. Twin working at the TESSA-3 spectrometer. States of a very particular nature were discovered in ^{152}Dy after the ^{108}Pd $(^{48}$Ca, $xn)$ $^{156-x}$Dy reaction. They form a very regular deformed rotor spectrum that is very rigid and stays visible over an extended region of spin values (Fig. 3.26). European research efforts using such spherical detectors include the EUROGAM and EUROBALL projects.

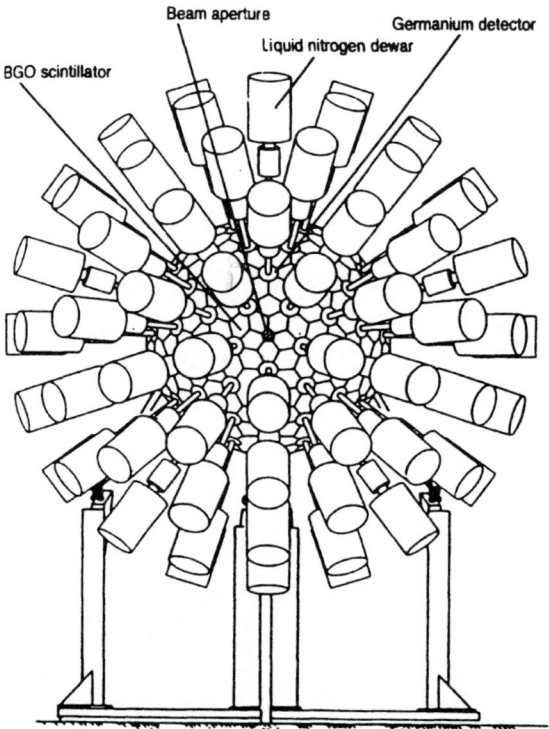

Fig. 3.25. Gammasphere, the powerful new nuclear gamma detector array constructed at Berkeley. The hollow spherical array of germanium and bismuth germanate crystals surrounds a heavy metal target that can be bombarded with ion beams from a cyclotron or van de Graaff accelerator. The array's inner and outer diameters are 50 and 90 cm. The liquid nitrogen dewars that cool the Ge detector elements extend the total diameter of the system to 2 m

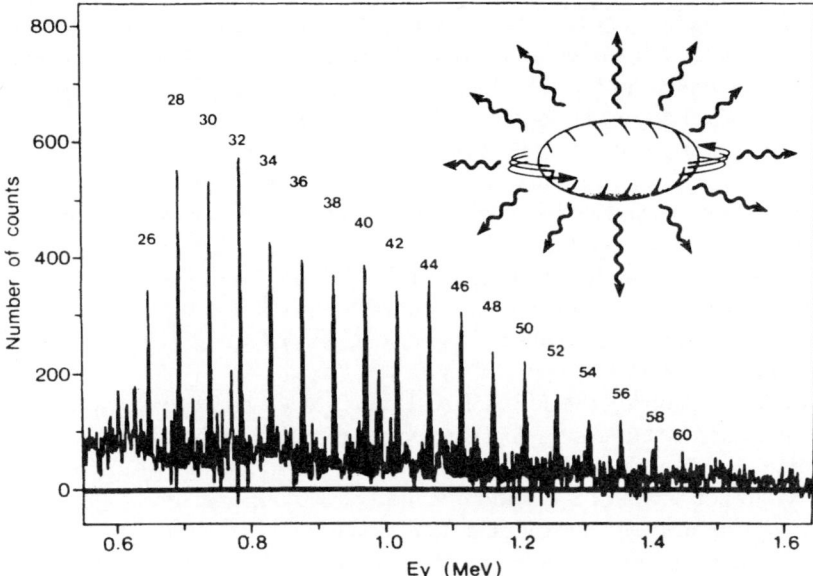

Fig. 3.26. The gamma emission spectrum, indicating the many quantum transitions linking the states in the superdeformed rotational band in ^{152}Dy. The numbers above each transition indicate the spin values of the corresponding band members. (Taken from NUPECC Report (1991) *Nuclear Physics in Europe: Opportunities and Perspectives* November, with permission)

Box V

EUROBALL: Probing the Rapidly Rotating Nucleus

At present, nuclear physicists are enjoying a most interesting and exciting period with the advent of the most recent generation of powerful gamma-ray spectrometers. In particular set-ups like EUROGAM (UK/France), GASP (Italy), GAMMASPHERE (USA) and, in the near future, EUROBALL, will be able to observe even the tiniest rotational motion of the atomic nucleus.

The total photopeak efficiency has gone up to about 10% and the gamma-ray detection efficiency has increased to 10^{-5} or better of the production cross section.

At present, at the Vivitron in Strasbourg, the newest implementation of EUROGAM, a full 4π gamma multidetector system contains up to 30 large-volume Compton-suppressed Ge detectors, up to 24 new polarization sensitive BGO shielded clovers to study how a highly excited nucleus in a state of very rapid rotation finds its way to the most stable configuration when cooled to a state having, in a quantum mechanical sense, the lowest possible angular momentum.

On July 1, 1994 an agreement was made between six European countries (Denmark, France, Germany, Italy, Sweden, and the UK) to construct

EUROBALL, the new gamma-ray spectrometer. This project will be the culmination of many years of collaborative efforts in this field of nuclear physics.

EUROBALL should become the most sensitive and sophisticated array of detectors. Its basic aim, as indicated before, is to understand how nucleons behave under extreme conditions of high excitation energy and rapid rotation. The newest array should be able to follow with unprecedented precision the way the nucleus dissipates its energy and rotational energy via gamma-ray emission.

The overall cost (about 20 million ECU) will be distributed between the various partners. The first places the apparatus will go to, at the anticipated date of finishing the array, will be Legnaro (near to Padua in Italy) and the CRN at Strasbourg in France.

The construction of the EUROBALL array with its unique capabilities will involve over a hundred scientists, engineers, and technical staff and it will provide a research facility not only for the participating teams but also for many outside users. It is a good example of cooperative efforts amongst a large number of laboratories to create a powerful and breathtaking facility on the threshold of the next millenium. It also proves that innovative technologies and forefront nuclear physics research can go hand-in-hand and can be supported in a combined effort over the borders of the contributing European countries.

Recent experiments in this field have now studied examples of nuclei in:

(i) the rare-earth region,
(ii) the Hg, Pb mass region and,
(iii) the medium-heavy mass A=80 and A=100 regions. Also the gamma transitions that connect these superdeformed structures to the nuclear structure in the slow rotation cold regime have been found and allow a good estimate of the excitation energy at which the superdeformed bands are formed.

Finally, one can even think of more exotic objects like hyperdeformed states with axes ratios of 3 : 1 (major to minor axis) for the axially symmetric case. Indications that such forms exist have been obtained very recently. Furthermore, the additions of such large amounts of angular momentum will inevitably push the nucleus to the limits beyond which it cannot sustain more angular momentum and cause fission or destabilization due to the very strong rotation.

3.4.3 Heating the Atomic Nucleus: Towards Chaotic Motion

Moving outside the region of cold nuclei by successively increasing the nuclear excitation energy, the regular nuclear structure (coherent types of nuclear collective motion like vibrational and rotational excitations) will, as a general

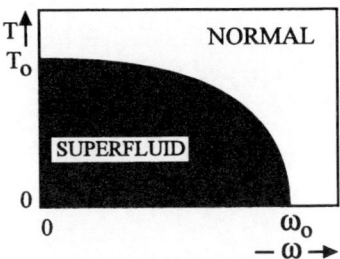

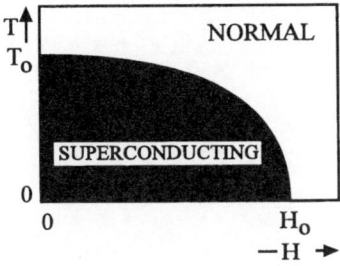

Fig. 3.27. Schematic illustration of the possible quenching of the nuclear superfluid phase inside the atomic nucleus by either increasing the internal temperature (T) and/or increasing the rotational frequency for the single-particle deformed potential. The nuclear phase diagram is compared with the corresponding superconducting electron gas in solid state physics (T, H axes)

rule, become "dissipated" through its coupling to the many intrinsic nucleonic excitations. The simple angular momentum 0^+ coupled pair correlations will become destroyed by the heating process. Looking only at the temperature and rotational types of degrees of freedom in the nuclear diagram, one sees, rather similar to the regular superconducting region on the (T, H) diagram (temperature, external magnetic field), that a superfluid region can be isolated at low excitation energies and small angular momenta (Fig. 3.27). By either making the nucleus rotate very rapidly and/or heating the nucleus intrinsically, one can leave this region of superfluidity where correlated 0^+ nucleon pairs dominate the properties of the interacting A-body system.

The general properties of such complex A-body systems can then, under certain conditions, start to exhibit chaotic features which are reflections of the basic interactions amongst the many nucleons inside the nucleus.

Before discussing some of the main features of this transition from more ordered motion, where symmetries in the interacting A-nucleon system dominate the nuclear structure properties, to more random distributions of nuclear excitations, we present a short technical box on statistical level distributions.

Box VI

Statistical Level Distributions

A property that is especially sensitive to the interacting A-nucleon system is the nuclear level spacing distribution or level densities.

We start from a simple 2×2 model Hamiltonian in which the elements of the matrix

$$\begin{pmatrix} H_{11} & H_{12} \\ H_{21} & H_{22} \end{pmatrix} , \tag{VI.1}$$

are randomly distributed (uncorrelated). We can then consider an ensemble of 2×2 matrices in which the probability for a given matrix is specified by some function $P(H)$ and which is given by the relation

$$P(H_{11}) = P(H_{22}) = P(H_{12}) = P(H_{21}) . \tag{VI.2}$$

One can then study the statistical properties of the eigenvalues and eigenvectors of this ensemble. If we call E_1 and E_2 the two eigenvalues, one can obtain (in an unnormalized form) a probability distribution for the relative energy difference, s, between the eigenvalues (with s defined as $s \equiv E_1 - E_2$)

$$P(s) = s \exp(-s^2) , \tag{VI.3}$$

which is referred to as the Wigner distribution (Fig. VI.1).

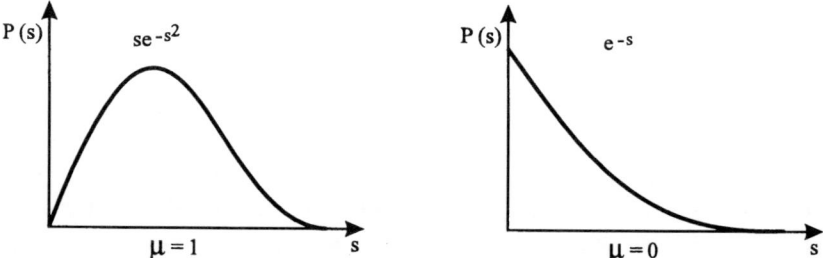

Fig. VI.1. Illustration of the two extreme distributions of energy eigenvalues: The Wigner probability distribution $s \exp(-s^2)$ corresponding to randomly distributed nuclear interaction matrix elements, and the Poisson distribution $\exp(-s)$ corresponding to a non-interacting system

This simple 2×2 matrix analysis illustrates the types of results that can be obtained starting from random matrix ensembles. However, in order to study more general properties of distriubtions of levels, one should extend the above method to properties of matrices with large dimensions. However, essentially the same Wigner distribution results. It is also said to correspond to the GOE or Gaussian-orthogonal ensemble of matrices (or interactions).

One can similarly show that in non-interacting systems (where, because of symmetries that govern the Hamiltonian describing the system, a number of vanishing H matrix elements appear), the energy spacing has a Poisson-like distribution (Fig. VI.1), and is given by the expression

$$P(s) = \exp(-s) . \tag{VI.4}$$

It is now possible to connect the above level spacing distributions via a semi-classical argument to classical ideas of integrable and non-integrable systems.

Rectangle   Circle

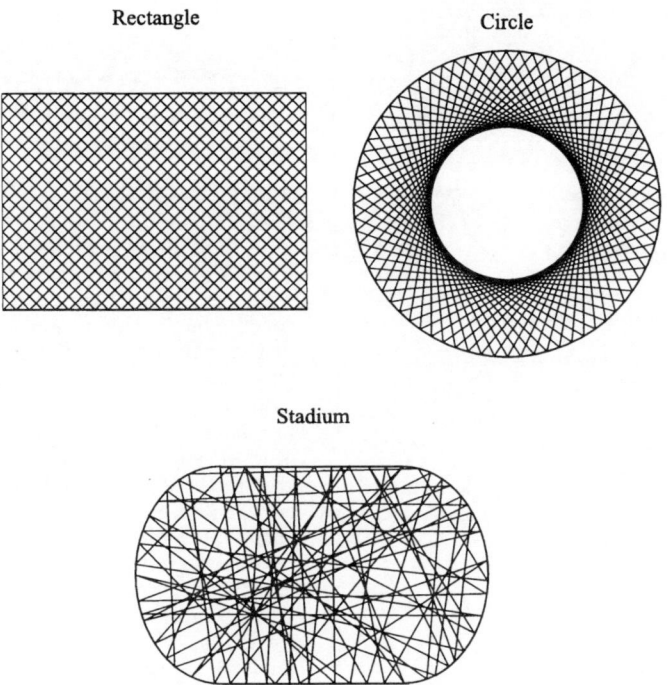

Stadium

Fig. VI.2. Some illustrative examples of domains (rectangle, region between two concentric circles) in which periodic, regular integrable motion can be obtained compared to a potential domain (stadium) in which random, non-integrable motion results

By considering a certain bounded region where classical motion is periodic (integrable), e.g., the rectangular domain or the domain spanned by two concentric circles (Fig. VI.2), or non-integrable (chaotic or partially chaotic), e.g., stadium domain or in the case of the well-known Sinai billard boundary conditions (Fig. VI.3), one can then regard the domain walls in a quantum-mechanical problem as the boundaries at which wavefunctions have to vanish. Solving these potential systems with a constant potential inside the domain and an infinite potential outside, one can find the eigenvalues and eigenstates and study the distributions of $P(s)$ where s is a measure of the distance between adjacent levels in the domain. For the classically integrable systems the resulting $P(s)$ distribution is Poisson-like and for the non-integrable systems where chaotic classical behavior can result, the Wigner or GOE distributions are found. Thus the answer to the question of whether, in the quantum case, spectral properties exist that reflect 'classical' chaotic motion can be given in the affirmative, albeit based on numerical evidence.

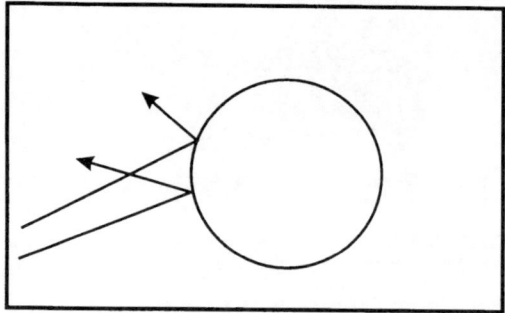

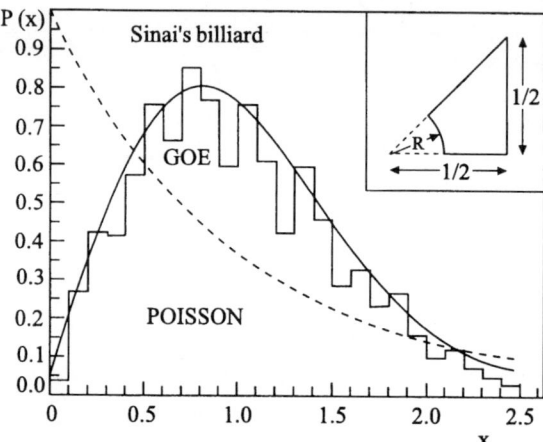

Fig. VI.3. The Sinai billard domain in which two, initally close trajectories diverge in time. This is due to the particular structure of the reflection on the circular part of the billard. The *lower part* of the figure shows the distribution of level separations [nearest-neighbor distribution $P(x)$] corresponding to the energy eigenvalues for the Sinai potential domain. It is evidently well described by a GOE (Gaussian Orthogonal Ensemble or Wigner distribution) distribution (Taken from H. Weidenmüller *Comm. Nuclear and Particle Physics* ©1986, Vol. 16, 199. Gordon & Breach, N.Y., with permission)

The above discussion justifies the interest in studying various spectral properties of atomic nuclei. One can certainly find experimentally interesting cases but the difficulty is that it is necessary to sample a sufficiently large ensemble of levels of the same angular momentum and parity in a given nucleus and study its nearest neighbor level distributions. Until recently the only information came from neutron and proton resonance studies. More recently, the situation has improved considerably through the availability of the "Nuclear Data Ensemble, NDE" which consists of all spin 1/2 s-wave neutron resonances measured by the Columbia group and of proton resonances measured by the group at TUNL (Fig. 3.28); the completely known spectrum of ^{26}Al between the ground state and proton threshold; spectroscopic data obtained via the extensive (n, γ) experiments over many nuclei; and also level distributions obtained starting from given model Hamiltonians in a theoretical way (e.g., within the nuclear shell model and also using the interacting boson model).

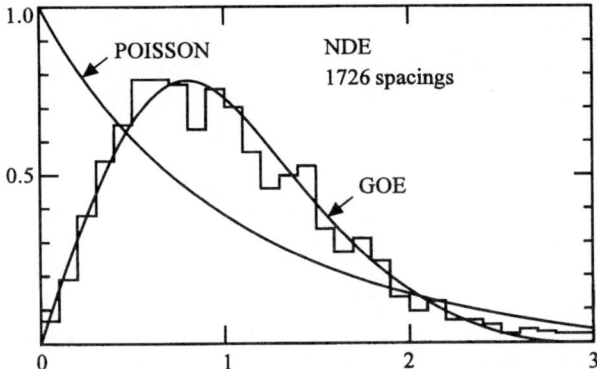

Fig. 3.28. Nearest-neighbor spacing distribution $P(x)$ corresponding to the Nuclear Data Ensemble (NDE) consisting of 1726 spacings. This NDE distribution follows a GOE distribution very closely. (Reprinted from A. Richter (1993) Nucl. Phys. A **553**, 417c. Elsevier Science, NL, with kind permission)

At high excitation energies it is still possible to observe simple excitation modes such as the electric giant dipole resonant state (GDR). This particular mode can be depicted as neutrons moving out-of-phase with protons. The resonance energy is related to the nuclear symmetry energy and the shape of the atomic nucleus. It has recently become possible to detect even states that correspond to the subsequent absorption of two photons of the correct energy, leading to two-phonon (multi-phonon?) states (Fig. 3.29).

By adding extra internal energy to a nucleus that exhibits a GDR, in particular the width of the resonant state increases with increasing excitation energy. Interesting modifications to an almost linear increase in width of the GDR with increasing excitation energy signals the energy (and thus also time) at which the given nucleus can no longer sustain coherent vibrations long enough to survive. Furthermore, the decay of GDR and the photons emitted during that process carry very interesting information about the time scale on which the resonance and the fission process decay.

If finally the "heating" of the nucleons inside the nucleus becomes of the order of the total binding energy of the nucleus, a global phase change from a nuclear fluid into a nuclear gaseous state may occur. Experimentally, today, the best way to map out this interesting transition zone is by using heavy ions in collision. With the advent, in particular, of RHIC (Relativistic Heavy-Ion Collider at Brookhaven National Laboratory) and a program to collide Pb with Pb ions at the LHC at CERN, completely new areas of physics will come under experimental scrutiny (see also the discussion in Chap. 5).

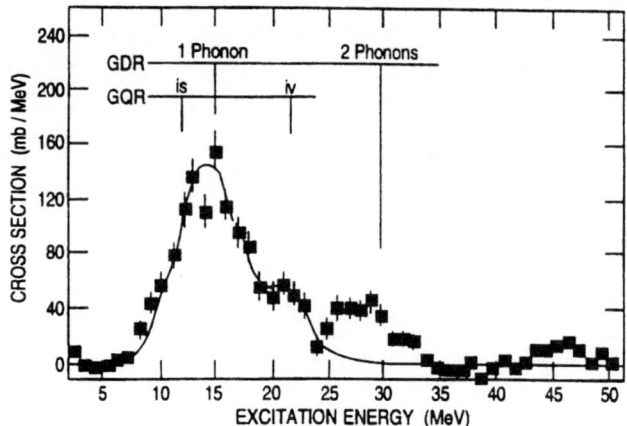

Fig. 3.29. Excitation spectrum of a Xe nucleus excited in a Xe–Pb collision at near relativistic energies. Besides the single-phonon, giant isoscalar dipole resonance (GDR-IS), double-phonon giant dipole resonance excitations are also clearly visible. The width of both the single- and double-GDR excitations illustrates the damping of these resonances by coupling to the internal modes of motion of intrinsic nucleonic excitations. (Taken from NSAC (1996) *Nuclear Science: A Long Range Plan* February, with kind permission)

3.4.4 Exotic Nuclei: Systems Far From Stability and Weakly Bound Quantum Systems

We now turn our attention to the direction in our three-dimensional space where the proton-to-neutron ratio (or, equivalently, the relative neutron excess $(N - Z)/A$) is the dominant variable. The most stable and thus best studied nuclei are located near to the region of beta stability. By adding more protons or neutrons, one is gradually leaving the bottom of the valley and thereby losing binding energy so that the system becomes unstable. It is the weak force (beta decay) that drives those unstable nuclei back towards the region of stability and fundamentally determines the mass landscape of atomic nuclei.

By extending the region of nuclei towards the extreme edges of stability, one reaches and then crosses the limits where adding one more neutron or proton leads to an unbound nucleus, i.e., one that is unstable with respect to one-nucleon emission. These limiting lines are called the drip lines and they present the ultimate borders of the nuclear landscape on a (N, Z) diagram (see fold-out chart).

With beams of particles impinging on a heavy nucleus, one can literally 'blow' a nucleus to pieces (spallation) and thus form a large range of nuclei in which the proton-to-neutron ratio (Z/N) is located far away from the region of beta stability. By then separating the so formed isotopes with the electric and magnetic fields of an isotope separator system, and bringing them rapidly to a measuring station one can study their nuclear properties within the short time for which such highly proton- or neutron-rich nuclei 'survive'. Such systems are operational at various places and are called ISOL (isotope separator on line) systems. A very well known separator, ISOLDE at CERN, has now found its place at the PS booster. The ISOLDE separator at CERN has been instrumental in discovering new nuclei extending to the ultimate limits, in particular for light nuclei, and is discussed in Box VII. An excerpt of the nuclear mass table for these light nuclei is given in Fig. 3.30 to illustrate the very exotic forms of nuclei that can and have been formed and studied over the last few years.

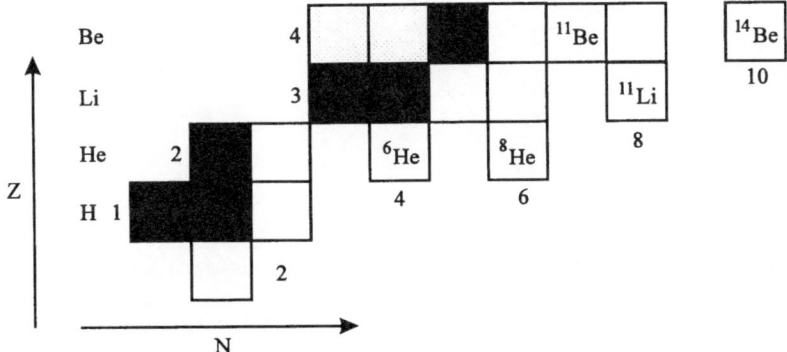

Fig. 3.30. Excerpt of the nuclear mass table for the very light nuclei. Stable nuclei are marked *black*. The very exotic and rapidly decaying halo nuclei ^{11}Li and ^{14}Be occur at the extreme edge of the neutron stability region

Box VII

The Exotic Beams of ISOLDE at CERN

ISOLDE, CERN's on-line isotope separator was originally installed at the 600 MeV SC (synchro-cyclotron). With the closure of this SC machine, the separator was moved to the PS booster in 1992 and was housed in a much larger building than before (Fig. VII.1). At present, the ISOLDE separator supplies a wide range of nuclear isotopes that are separated in charge and mass and serve a large number of experiments located in this new building. The experimental program encompasses nuclear physics, fundamental tests of the standard theory of electroweak interactions, and also covers research topics in solid state physics and biomedical research. Besides the large range of experiments running at ISOLDE, a pilot experiment using radioactive beams at ISOLDE (called REX-ISOLDE) has been approved and will be installed in an enlarged experimental hall. This post-accelerator project will extend the already impressive experimental facilities. In the figure, the planned lay-out of the enlarged experimental hall is shown.

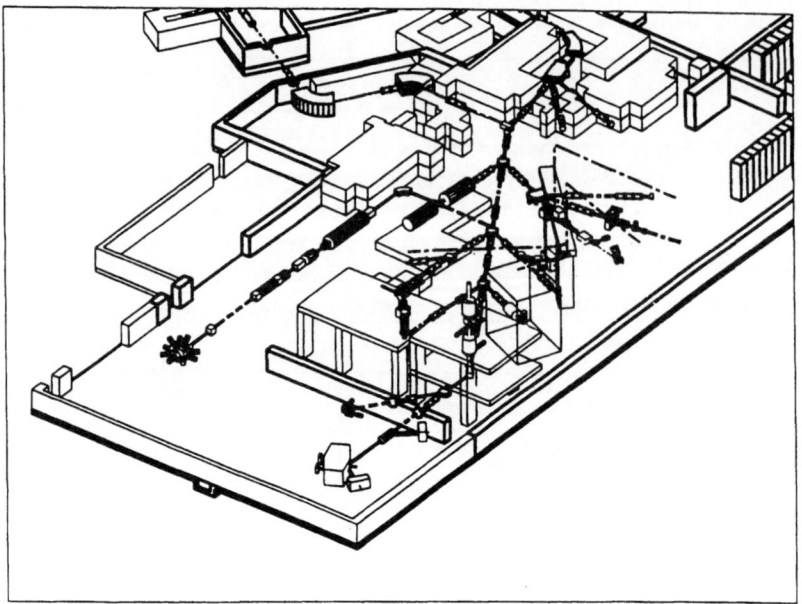

Fig. VII.1. Planned extension to the experimental hall of the ISOLDE on-line separator served by CERN's PS-booster accelerator. On the *left* in the extension is the beam-line for the REX-ISOLDE new project for studying exotic neutron-rich $A \simeq 50$–60 nuclei. *Behind* this is the existing ISOLDE hall, housing a wide variety of experiments. (Taken from CERN Courier (1995) December, with permission)

The success of ISOLDE has been built largely on state-of-the-art skills enabling the development of new target and beam techniques on site, together with the ever-increasing need for beams of exotic nuclei at the extreme limits of nuclear stability. In particular, in satisfying the need to reach the stability limits, acceleration of the radioactive elements created to higher energies (2–6 MeV/nucleon) is a topic of very high priority. This has been underlined in discussion meetings and is supported in large part by the Nuclear Physics European Collaboration Committee (NUPECC).

The REX-ISOLDE project at CERN is a collaboration between various European groups. The radioactive elements will be accelerated by a radio-frequency quadrupole (RFQ) and a linac. In order to optimize the beam properties, the radioactive nuclei formed at the PS booster will be collected in a Penning trap which will also serve as a beam buncher. Experience with Penning traps at ISOLDE has been gained by the Mainz group (H. J. Kluge, G. Bollen, et al.) and resulted in unique mass measurements for highly unstable nuclei. With this setup it will be possible to form neutron-rich light nuclei in a first set of experiments intended to test the nuclear shell model and its extrapolations towards the neutron drip line and to look for new regions of deformation.

The new technical possibility to accelerate even short-lived unstable isotopes is allowing one to approach ever closer to the drip-lines and to study atomic nuclei at the most extreme ratios of protons to neutrons possible. In those regions of the nuclear mass table, it may well be that some of the more standard ideas about how nucleons behave in the atomic nucleus have to be modified, or even more drastic, be rejected completely. For light nuclei one can reach regions where the number of neutrons compared to protons is indeed overwhelming. Nuclei like ^{8}He, ^{11}Li, and ^{14}Be are examples and have indeed presented fully unexpected results to the scientific community.

The atomic nucleus ^{10}Li decays almost immediately, whereas ^{11}Li, containing one neutron more, is just bound with a tiny binding energy of 250 keV. One thus encounters systems where an A nucleon system does not exist but the $A + 1$ system does! The very weak binding energy, according to basic quantum mechanics will cause the system to be formed in a state where the barely bound nucleons will, on average, move very far out from the center of the nucleus. These very long tails of the nuclear wavefunctions lead to what has been called 'halos'. In the case of ^{11}Li, the outer two neutrons form a 'halo' system as illustrated in Fig. 3.31. A very direct proof of the subdivision into tightly bound and strongly localized nucleons on the one hand and a halo (loosely bound system) outside is found in the momentum distributions of nucleons moving away from the final nucleus. According to the Heisenberg uncertainty principle, the core nucleons should have a very large momentum spread whereas the halo neutrons should correspond to a relatively well-defined momentum. This has been confirmed experimentally and is illustrated in Fig. 3.32.We refer to a recent review article on halo structures in nuclei by Tanihata [3.168] for all technical details.

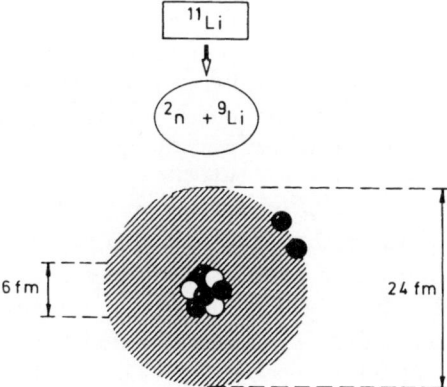

Fig. 3.31. Schematic drawing of the halo nucleus ^{11}Li in which 3 protons and 6 neutrons form a fairly well closed core nucleus ^{9}Li with 2 extra neutrons, loosely bound to the inner core and extending towards large radial distances ($\simeq$ 24 fm) thereby forming a neutron halo

For nuclei possessing neutron halos, and also the less exotic heavy nuclei with neutron skins, various new modes of oscillation become possible in which the core moves relative to the outer neutrons and unusual types of vibrations

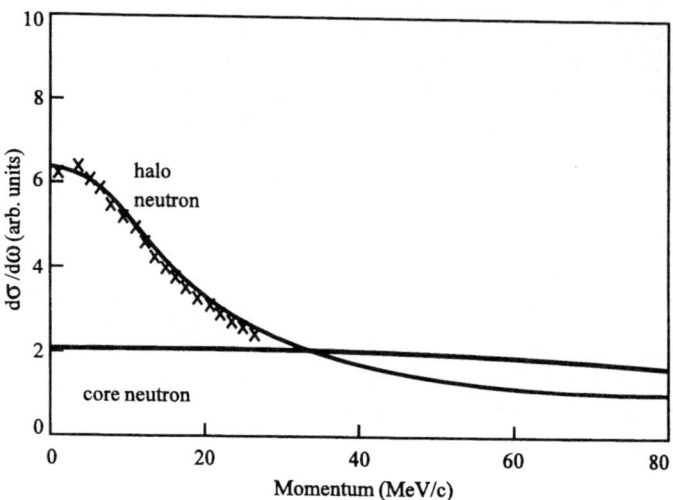

Fig. 3.32. Momentum distribution of a core neutron (*flat curve*) and of a loosely bound halo neutron (*upper curve*) in the nucleus ^{11}Be, measured with respect to the final nucleus after nuclear fragmentation at high energies. These curves are consistent with the spatial distribution of the corresponding neutron and satisfy the Heisenberg uncertainty principle. (Taken from NSAC (1996) *Nuclear Science: A Long Range Plan* February, with kind permission)

and rotations occur. These modes of nuclear motion have hardly been investigated and will form part of extensive research programs at facilities for radioactive ion beams that are planned or under construction.

The opposite side of the valley of stability, the region where proton excess is to be expected, can be studied using the radioactive-ion-beam facilities right through to the heaviest Pb region of nuclei. Because the protons carry a charge, an outer Coulomb barrier allows very weakly bound systems to survive for quite a few seconds with protons almost literally 'dripping' through the Coulomb barrier. The lifetimes of these extreme nuclei with respect to proton emission will be quite sensitive to the underlying dynamics of the forces within the system and are at best poorly understood at present. In these medium-heavy and heavy proton-rich nuclei it is possible to try to follow the particular line of nuclei with equal number of protons and neutrons, i.e., $N = Z$. A new form of correlation, called proton–neutron superfluidity, may show up but still remains to be confirmed experimentally. Attempts to extend this $N = Z$ line have recently culminated in the discovery of the heaviest $N = Z$ nucleus so far, namely ^{100}Sn, which for a long time has been the 'holy grail' of nuclear physics, both at the French institute GANIL and at the German GSI facility. Experiments in this region are rapidly extending our knowledge which was previously based only on extrapolations, a procedure that has often proved inadequate when entering 'terra incognita'.

The theoretical understanding of such highly unstable proton-rich or neutron-rich systems is still in its infancy. As just mentioned, extrapolating out of the valley of beta stability, where the nuclear structure is dominated by single-particle motion in an average field created by all other nucleons together with a strong spin–orbit coupling, may well prove to be the wrong approach when seeking reliable results. Near the neutron drip line it may well be that the pair scatter of neutrons from the barely bound orbitals into unbound scattering states will modify the underlying organization of nucleons inside the nucleus in a major way. Some results of extrapolations, based on state-of-the-art Hartree–Fock–Bogoliubov methods and calculations are illustrated in Figs. 3.33 and 3.34).

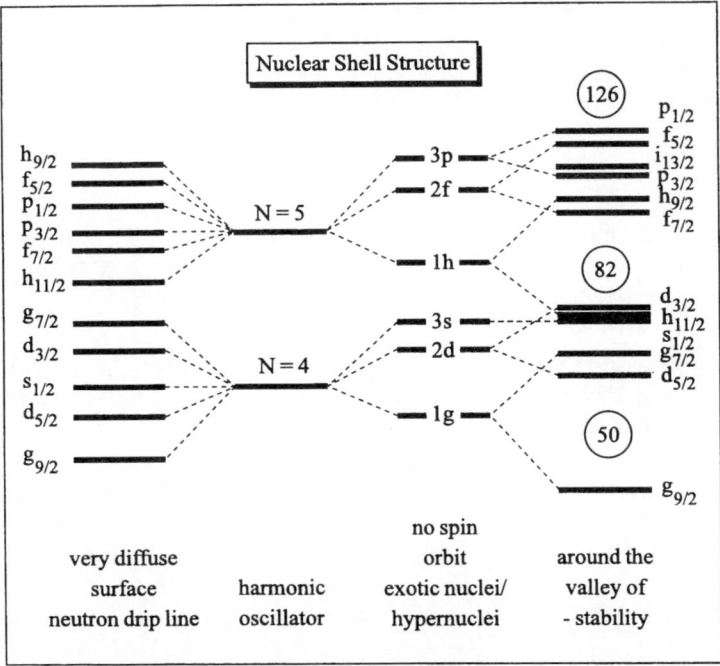

Fig. 3.33. Nuclear single-particle ordering in various average fields. The levels are characterized by the quantum numbers N, l, j. At the far *left* is the structure in a potential with a spin–orbit term but with vanishing $\hat{l}^2$ term (corresponding to a rather diffuse nuclear surface). *Next*, the fully degenerate spherical harmonic oscillator spectrum (for $N = 4$ and $N = 5$) is given; *then* the level structure for a harmonic oscillator plus a $\hat{l}^2$ term with vanishing spin–orbit term. On the far *right* is the nuclear single-particle structure, typical for stable nuclei around the region of beta stability. (Taken from NSAC (1996) *Nuclear Science: A Long Range Plan* February, with kind permission)

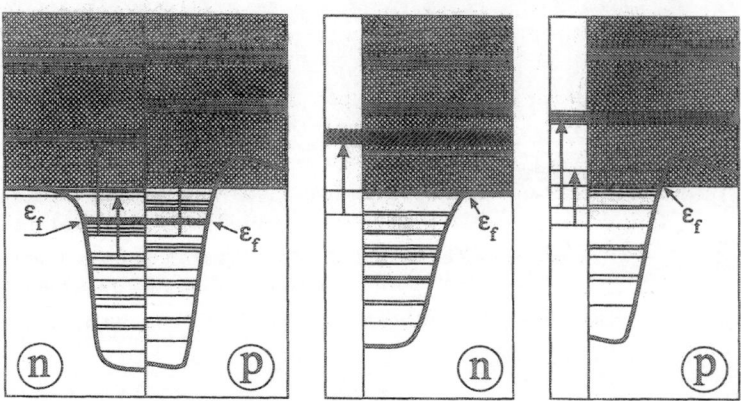

Fig. 3.34. Typical energy spectra (*left*) near the region of beta stability, (*center*) near to the neutron drip-line region, and (*right*) near to the proton drip line. The illustrative nuclei are ^{120}Sn, ^{150}Sn and ^{100}Sn, respectively. (Adapted from W. Nazarewicz et al. (1994) Phys. Rev. C **50**, 2860. American Physical Society, with permission)

The creation and, more importantly, the acceleration of beams of radioactive ions now enables one to study certain reactions that occur at the 'heart' of stars and thus are vital in our understanding of how elements have formed in nucleosynthesis and at what rate (energy production and cooling of stars). These points will be discussed further in Chap.7.

During the past 20 years, in trying to form nuclei with high proton-neutron ratios, a special route has been explored towards the region of superheavy nuclei. This promised 'island' near to the magic numbers $Z = 114$ and $N = 170 - 180$ is getting closer thanks to ingenious experimentation in particular at the GSI by a team headed by Armbruster and Munzenberg. At present, elements of $Z = 107$ up to 112 have been synthesized and their shear existence shows that nuclear stabilizing shell effects are strong enough to counterbalance the liquid drop fission channel. In Fig. 3.35, we show the present limits of the synthesized elements where the road seems to stop or, as often drawn in a cartoon-type of way, falls into the sea of instability. The technique of cold-fusion is already exploring the next element of $Z = 110$ through the reaction ^{208}Pb(^{62}Ni,n)269(?)110 with an expected production rate of about 1 atom for each 800 h of machine operations! Needless to say, here one is indeed reaching the extreme limits of stability. An extremely readable account by Armbruster and Munzenberg describing the road towards creating superheavy elements appeared a few years ago in Scientific American May, 1989 [3.155].

Before closing this section, and still keeping to the theme of exotic nuclei, we remark that the region of exceptional structures subject to the strong nuclear binding forces is surely even more rich than we can even anticipate at present. In mapping out an expanded nuclear mass table where we put

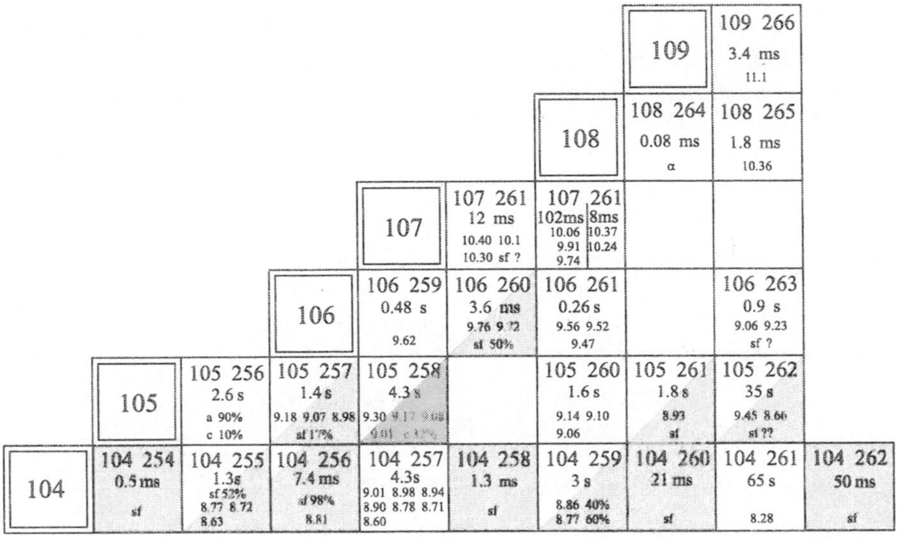

											109	
											109 266 3.4 ms 11.1	
										108	**108 264** 0.08 ms α	**108 265** 1.8 ms 10.36
						107	**107 261** 12 ms 10.40 10.1 10.30 sf ?	**107 261** 102ms 8ms 10.06 10.37 9.91 10.24 9.74				
				106	**106 259** 0.48 s 9.62	**106 260** 3.6 ms 9.76 9.72 sf 50%	**106 261** 0.26 s 9.56 9.52 9.47				**106 263** 0.9 s 9.06 9.23 sf ?	
		105	**105 256** 2.6 s a 90% c 10%	**105 257** 1.4 s 9.18 9.07 8.98 sf 17%	**105 258** 4.3 s 9.30 9.17 9.08 9.01 c 19%		**105 260** 1.6 s 9.14 9.10 9.06	**105 261** 1.8 s 8.93 sf	**105 262** 35 s 9.45 8.66 sf ??			
104	**104 254** 0.5 ms sf	**104 255** 1.3 s sf 52% 8.77 8.72 8.63	**104 256** 7.4 ms sf 98% 8.81	**104 257** 4.3 s 9.01 8.98 8.94 8.90 8.78 8.71 8.60	**104 258** 1.3 ms sf	**104 259** 3 s 8.86 40% 8.77 60%	**104 260** 21 ms sf	**104 261** 65 s 8.28	**104 262** 50 ms sf			

Fig. 3.35. Excerpt of the nuclear mass table in the region of transactinide nuclei at the highest Z-values synthesized as yet. The various decay modes: alpha-decay (*open white*), electron capture (*dark grey*) and spontaneous fission (*light grey*) are indicated. (Reprinted from A. Richter (1993) Nucl. Phys. A **553**, 417c. Elsevier Science, NL, with kind permission)

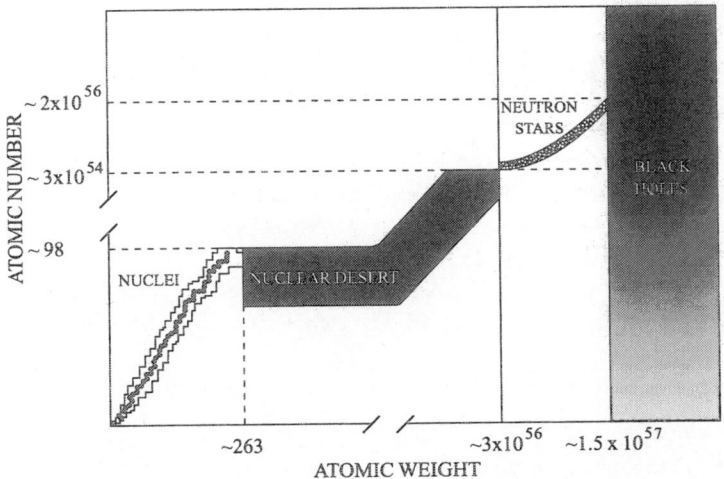

Fig. 3.36. An extreme form of the chart of nuclei showing all known forms of stable matter. Between the region of the heaviest atomic elements ($A \simeq 260 - 270$) and neutron stars ($A \simeq 10^{56} - 10^{57}$), a vast nuclear 'desert' may exist or, alternatively, a region where strange forms of quark matter can be found. (Adapted from H.J. Crawford and C.H. Greiner ©1994 *Scientific American* January, with permission)

all known forms of stable nuclear matter (Fig. 3.36), one observes a vast unknown unpopulated nuclear desert between known normal nuclei up to mass $A \simeq 260$ and the neutron stars that may, to a first approximation (see also Sect. 3.1) be treated as giant nuclei with atomic weights of about $10^{+56} - 10^{+57}$.

3.5 Further Reading

Before listing some textbooks discussing nuclear structure, we cite two more popular review papers:

3.1 Hodgson, P.E. (1994) Contemp. Phys. **35**, 329
3.2 Wilkinson, D. (1984) Nucl. Phys. **A421**, 1c

An overview of technical developments in the 1990s, for advanced readers:

3.3 Johnson, N.R. (ed.) (1990) Nuclear Structure in the Nineties, Nucl. Phys. **A520**

A large number of textbooks are devoted to nuclear structure. We first give some that concentrate on the topic as a whole:

3.4 Bohr, A., Mottelson, B. (1969) *Nuclear Structure*, Vol. 1 (Benjamin, New York)
3.5 Bohr, A., Mottelson, B. (1975) *Nuclear Structure*, Vol. 2 (Benjamin, New York)
3.6 Eisenberg, J.M., Greiner, W. (1987) *Nuclear Models*, Vol. 1, 3rd ed. (North-Holland, Amsterdam)
3.7 Eisenberg, J.M., Greiner, W. (1987) *Excitation Mechanisms of the Nucleus*, Vol. 2, 3rd ed. (North-Holland, Amsterdam)
3.8 Eisenberg, J.M., Greiner, W. (1976) *Microscopic Theory of the Nucleus*, Vol. 3 (North-Holland, Amsterdam)
3.9 Hornyack, W.F. (1975) *Nuclear Structure* (Academic, New York)
3.10 Ring, P., Schuck, P. (1980) *The Nuclear Many-body Problem* (Springer, Berlin Heidelberg)
3.11 de Shalit, A., Fesbach, H. (1974) *Theoretical Nuclear Physics* (Wiley, New York)

Textbooks concentrating more on shell model methods:

3.12 Brussaard, P.J., Glaudemans, P.W.M. (1977) *Shell Model Applications in Nuclear Spectroscopy* (North-Holland, Amsterdam)
3.13 Heyde, K.L.G. (1994) *The Nuclear Shell Model* Study edition, 2nd ed. (Springer, Berlin Heidelberg)
3.14 Lawson, R.D. (1980) *Theory of the Nuclear Shell Model* (Clarendon Press, Oxford)

3.15 Mayer, M.G., Jensen, H.D. (1955) *Elementary Theory of Nuclear Shell Structure* (Wiley, New York)

3.16 de Shalit,A., Talmi, I. (1963) *Nuclear Shell Theory* (Academic, New York)

3.17 Talmi, I. (1993) *Simple Models of Complex Nuclei: The Shell Model and Interacting Boson Model* (Harwood, New York)

Those textbooks that deal with collective motion in the nucleus include, of course, the monumental volume series of Bohr and Mottelson mentioned above. Some further references are:

3.18 Kumar, K. (1984) *Nuclear Models and the Search for Unity in Nuclear Physics* (Universitetsforlanger, Oslo)

3.19 Nilsson, S.G., Ragnarsson, I. (1995) *Nuclear Shells and Shapes* (Cambridge University Press, New York)

3.20 Rowe, D.J. (1970) *Nuclear Collective Motion* (Methuen, New York)

Two textbooks discussing numerical methods and computer codes that describe nuclear structure and nuclear reaction properties are:

3.21 Langanke, K., Maruhn, J.A., Koonin, S.E. (ed.) (1991) Computational Nuclear Physics 1: Nuclear Structure (Springer New York)

3.22 Langanke, K., Maruhn, J.A., Koonin, S.E. (ed.) (1993) Computational Nuclear Physics 2: Nuclear Reactions (Springer New York)

In studying various nuclear structure processes, one first has to understand how the nucleon interactions generate an average field. On Hartree–Fock theory, the details are well described in the general nuclear structure textbooks. A number of more detailed but important papers in this field, as well a number of review papers on self-consistent Hartree–Fock techniques with applications, are the following:

3.23 Baktash, C., Haas, B., Nazarewicz, W. (1995) Ann. Rev. Nucl. Part. Sci. **45**, 485

3.24 Aberg, S., Flocard, H., Nazarewicz, W. (1990) Ann. Rev. Nucl. Part. Sci. **40**, 439

3.25 Goodman, A.L. (1979) Adv. Nucl. Phys. **11**, 283

3.26 Quentin, P., Flocard, H. (1978) Ann. Rev. Nucl. Sci. **28**, 523

3.27 Skyrme, T.H.R. (1956) Phil. Mag. **1**, 1043

3.28 Vautherin, D., Brink, D.M. (1972) Phys. Rev. **C5**, 626

3.29 Waroquier, M., Heyde, K., Wenes, G. (1983) Nucl. Phys. **A404**, 269

3.30 Waroquier, M., Ryckebusch, J., Moreau, J., Heyde, K., Blasi, N., Van der Werf, S., Wenes, G. (1987) Phys. Rep. **148**, 249

A basic article combining Hartree–Fock methods with the time dependence of nuclear dynamics is

3.31 Bonche, P., Koonin, S., Negele, J.W. (1976) Phys. Rev. **C13**, 1226

Liquid drop concepts are described by Bethe and Bacher, and von Weizsäcker but more recent parametrizations have also been made; they can be compared with extensive mass compilations:

3.32 Audi, G., Wapstra, A.H. (1995) Nucl. Phys. **A595**, 409
3.33 Bethe, H.A., Bacher, R.F. (1936) Rev. Mod. Phys. **8**, 82
3.34 Mattauch, H.E., Thiele W., Wapstra, A.H. (1965) Nucl. Phys. **67**, 1
3.35 Moller, P., Nix, J.R., Myers, W.D., Swiatecki, W.J. (1995) At. Data Nucl. Data Tables **59**, 185
3.36 Myers, W.D., Swiatecki, W.J. (1966) Nucl. Phys. **81**, 1
3.37 Weizsäcker, von C.F. (1935) Z.Phys. **96**, 431

Concerning the shell model, a large body of references exists relating to aspects of constructing effective interactions starting from the bare nucleon interactions, the construction of effective forces in given model spaces, on large-scale shell model studies, and recent developments concerning Shell Model Monte-Carlo extensions. We give here a number of important references in this domain.

3.38 Barrett, B.R., Kirson, M.W. (1973) Adv. Nucl. Phys. **6**, 219
3.39 Brandow, B.H. (1967) Rev. Mod. Phys. **39**, 771
3.40 Brown, G.E., Kuo, T.T.S. (1967) Nucl. Phys. **A92**, 481
3.41 Brueckner, K.A., Eden, R.J., Francis, N.C. (1955) Phys. Rev. **100**, 891
3.42 Brueckner, K.A., Gammel, J.L. (1958) Phys. Rev. **109**, 1023
3.43 Hjorth-Jensen, M., Osnes, E., Müther, H. (1992) Ann. Phys. **213**, 102
3.44 Hjorth-Jensen, M., Kuo, T.T.S., Osnes, E. (1995) Phys. Rep. **261**, 125
3.45 Kuo, T.T.S. (1981) in *Topics in Nuclear Physics*, Lect. Notes Phys., Vol.144 (Springer, Berlin Heidelberg)
3.46 Kuo, T.T.S., Brown, G.E. (1966) Nucl. Phys. **85**, 40
3.47 Schucan, T.H., Weidenmüller, H.A. (1972) Ann. Phys. (N.Y.) **73**, 108
3.48 Schucan, T.H., Weidenmüller, H.A. (1973) Ann. Phys. (N.Y.) **76**, 425

Discussions of large-scale shell model calculations; the codes (acronyms of the various large-scale shell model codes are given) and applications include:

3.49 Brown, B.A., Wildenthal, B.H. (1988) Ann. Rev. Nucl. Sci. **38**, 29
3.50 Caurier, E. (1989) ANTOINE CRN Strasbourg
3.51 Caurier, E., Zuker, A.P., Poves, A., Martinez-Pinedo, G. (1994) Phys. Rev. **C50**, 225
3.52 van Hees, A.G.M., Glaudemans, P.W.M. (1981) RITSCHIL Z. Phys. **A303**, 267
3.53 McRae, W.D., Etchegoyen, A., Brown, B.A. (1988) OXBASH MSU Report 524

3.54 Nakada, K., Sebe, T., Otsuka, T. (1994) Nucl. Phys. **A571**, 467
3.55 Schmid, K.W., Grümmer, F. (1987) VAMPIR Rep. Progr. Phys. **50**, 731
3.56 Sebe, T., Harvey, M. (1968) AECL report No. 3007
3.57 Wildenthal B.H. (1976) Varenna Lectures **69**, 383
3.58 Wildenthal, B.H., McGrory, J.B., Halbert, E.C., Graber, H.D. (1971) Phys. Rev. **C4**, 1708

A recent conference proceedings on state-of-the-art shell model studies is:

3.59 Wyss, R. (ed.) (1995) Proc. Int. Symp. on New Nuclear Structure Phenomena in the Vicinity of Closed Shells, Phys. Script. T56

References on Shell Model Monte-Carlo calculations:

3.60 Honma, M., Mizusaki, T., Otsuka, T. (1996) Phys. Rev. Lett. **77**, 3315
3.61 Koonin, S.E., Dean, D.J., Langanke, K. (1997) Phys. Rep. **278**, 1
3.62 Lang, G.H., Johnson, C.W., Koonin, S.E., Ormand, W.E. (1993) Phys. Rev. **C48**, 1518
3.63 Mizusaki, T., Honma, M., Otsuka, T. (1996) Phys. Rev. **C53**, 2786
3.64 Ormand, W.E., Dean, D.J., Johnson, C.W., Lang, G.H., Koonin, S.E. (1994) Phys. Rev. **C49**, 1422
3.65 Pudliner, B.S., Pandharipande, V.R., Carlson, J., Wiringa, R.B. (1995) Phys. Rev. Lett. **74**, 4396

Works discussing the importance of possible low-lying excitations from outside the regular shell model spaces, with many references to the vast literature, are:

3.66 Heyde, K., Van Isacker, P., Waroquier, M., Wood, J.L., Meyer, R.A. (1983) Phys. Rep. **102**, 291
3.67 Wood, J.L., Heyde, K., Nazarewicz, W., Huyse, M., Van Duppen, P. (1992) Phys. Rep. **215**, 101

There is a very large literature on symmetries in nuclear physics. First we give a couple of more general introductory references:

3.68 Gross, D.J. (1995) Physics Today December, p. 46
3.69 Peierls, R. (1992) Contemp. Physics **33**, 221
3.70 Rowe, D.J., Nash, C. (1991) Symmetry, Art and Nuclear Physics (University of Toronto, Toronto)
3.71 Wambach, J. (1991) Contemp. Phys. **32**, 291
3.72 Yang, C.N. (1991) AAPPSB Bull. Vol. 1, No. 3, p. 3

Concepts relating to structural symmetries, and the relation between geometrical symmetries and various modes of motion are extensively discussed in the works of Bohr and Mottelson (Vol. 2) and the volumes of Eisenberg and Greiner. Here we add a couple of references on the 'scissors' mode:

3.73 Berg, U.E.P., Kneissl, U. (1987) Ann. Rev. Nucl. Part. Sci. **37**, 33

3.74 Bohle, D., Richter, A., Steffen, W., Dieperink, A.E.L., Lo Iudice, N., Palumbo, F., Scholten, O. (1984) Phys. Lett. **B137**, 27

3.75 de Coster, C., Heyde, K., Richter, A., Wörtche, H.J. (1992) Nucl.Phys. **A542**, 375

3.76 Kneissl, U., Pitz, H.H., Zilges, A. (1996) Prog. Part. Nucl. Phys. **37**, 349

3.77 Richter, A. (1996) in *Building Blocks of Nuclear Structure*, ed. by A. Covello (World Scientific, Singapore)

3.78 Ziegler, W., Rangacharyulu, C., Richter, A., Spieler, C. (1990) Phys. Rev. Lett. **65**, 2515

The history of symmetry concepts in nuclear physics is impressive and here we give a chronological list of some keynote articles

3.79 Heisenberg, W. (1932) Z.Phys. **77**, 1

3.80 Wigner, E.P. (1937) Phys. Rev. **51**, 106

3.81 Racah, G. (1943): Phys. Rev. **63**, 367

3.82 Mayer, M.G. (1949) Phys. Rev. **75**, 1969

3.83 Haxel, O., Jensen, H.H.D., Suess, H.E. (1949) Phys. Rev. **75**, 1766

3.84 Bohr, A. (1951) Phys. Rev. **81**, 134

3.85 Bohr, A. (1952) Mat. Fys. Medd. Dan. Vid. Selsk. **26**, No. 4

3.86 Bohr, A., Mottelson, B. (1953) Mat. Fys. Medd. Dan. Vid. Selsk. **27**, No. 16

3.87 Elliott, J.P. (1958) Proc. Roy. Soc. **A245**, 128, 562

3.88 Elliott, J.P., Harvey, M. (1963) Proc. Roy. Soc. **A272**, 557

3.89 Arima, A, Iachello, F. (1975) Phys. Lett. **B57**, 39

3.90 Arima, A., Iachello, F. (1975) Phys. Rev. Lett. **35**, 1069

3.91 Rowe, D.J. (1996) Prog. Part. Nucl. Phys. **37**, 265

On dynamical symmetries there is also a large amount of literature and, for mathematical aspects, we refer to some general textbooks:

3.92 Gilmore, B. (1974) *Lie Groups, Lie Algebras and some of their Applications* (Wiley, New York)

3.93 Hamermesh, M. (1962) *Group Theory and its Applications to Physical Problems* (Addison-Wesley, Reading, MA)

3.94 Iachello, F. (1983) *Lie Groups, Lie Algebras and Some Applications*, Trento Lecture Notes, UT. F97

3.95 Lipkin, H. (1965) *Lie Groups for Pedestrians* (North-Holland, Amsterdam)

3.96 Parikh, J.C. (1978) *Group Symmetries in Nuclear Structure* (Plenum, New York)

3.97 Wybourne, B.G. (1974) *Classical Groups for Physicists* (Wiley, New York)

A very extensive literature deals with the interacting boson model (IBM). Here we quote a number of essential textbooks, the original Ann. Phys. (N.Y.) articles, and some more general articles written in a popular style:

3.98 Casten, R.F. (ed.) (1993) *Algebraic Approaches to Nuclear Structure* (Gordon & Breach, New York)

3.99 Iachello, F., Arima, A. (1988) *The Interacting Boson Model* (University Press, Cambridge)

3.100 Iachello, F., van Isacker, P. (1991) *The Interacting Boson-Fermion Model* (University Press, Cambridge)

3.101 Arima, A., Iachello, F. (1976) Ann. Phys. (N.Y.) **99**, 253

3.102 Arima, A., Iachello, F. (1978) Ann. Phys. (N.Y.) **111**, 201

3.103 Arima, A., Iachello, F. (1979) Ann. Phys. (N.Y.) **123**, 436

3.104 Casten, R.F. (1984) Comments Nucl. Part. Phys. **12**, 119

3.105 Dieperink, A.E.L. (1985) Comments Nucl. Part. Phys. **14**, 25

3.106 Talmi, I. (1983) Comments Nucl. Part. Phys. **11**, 241

Rotational properties of deformed nuclei have quite a long history. We start by listing chronologically some of the original older papers:

3.107 Nilsson, S.G. (1955) K. Dan. Vidensk. Selsk. Mat. Fys. Medd. **29**, No. 16

3.108 Strutinski, V.M. (1967) Nucl. Phys. **A95**, 420

3.109 Strutinski, V.M. (1968) Nucl. Phys. **A122**, 1

3.110 Brack, M., Damgaard, J., Jensen, A.S., Pauli, H.C., Strutinsky, V.M., Wong, C.Y. (1972) Rev. Mod. Phys. **44**, 320

These concepts are also discussed at length in the books of Bohr and Mottelson, Eisenberg and Greiner, Nilsson and Ragnarsson and Ring and Schuck. Some older papers concentrating on deformation, backbending and also presenting extensive data are:

3.111 Bunker, M.E., Reich, C.W. (1971) Rev. Mod. Phys. **43**, 438

3.112 Johnson, A., Szymanski, Z. (1973) Phys. Rep. **7**, 181

3.113 Ogle, W., Wahlborn, S., Piepenbring, R., Frederiksson, S. (1971) Rev. Mod. Phys. **43**, 424

3.114 Ragnarsson, I., Nilsson, S.G., Sheline, R.K. (1978) Phys. Rep. **45**, 1

Some recent review papers concentrating on many aspects of rapidly rotating and strongly deformed nuclei, e.g., superdeformation and hyperdeformation, including the original discovery paper by Twin et al.:

3.115 Aberg, S., Flocard, H., Nazarewicz, W. (1990) Ann. Rev. Nucl. Part. Sci. **40**, 439

3.116 Beausang, C.W., Simpson, J. (1996) J. Phys. G **22**, 527

3.117 Firestone, R.B., Singh, B. (1995) Table of Superdeformed Nuclear Bands and Fission Isomers LBL-35916, UC-413

3.118 Garrett, J.D. (1988) in *The Response of Nuclei under Extreme Conditions*, ed. by Broglia, R.A., Bertsch, G. (Plenum, New York)

3.119 Janssens, R.V.F., Khoo, T.L. (1991) Ann. Rev. Nucl. Part. Sci. **41**, 321

3.120 LaFosse, D.R. et al. (1996) Phys. Rev. **C54**, 1585

3.121 Nolan, P.J., Twin, P. (1988) Ann. Rev. Nucl. Part. Sci. **38**, 533
3.122 Sharpey-Schafer, J.F., Simpson, J. (1988) Rep. Prog. Phys. **21**, 293
3.123 Twin, P. et al. (1986) Phys. Rev. Lett. **57**, 811

We also include a number of more popular articles discussing properties of rapidly rotating nuclei and the experimental facilities to detect them:

3.124 Garrett, J. (1984) Comm. Nucl. Part. Phys. **13**, 1
3.125 Goldhaber, J. (1991) LBL Research Review, Spring, p. 22
3.126 Hellemans, A. (1996) Science **271**, 24
3.127 Lieb, K.P., Broglia, R., Twin, P. (1994) Nucl. Phys. News **3**, 21
3.128 Newton, J.O. (1989) Contemp. Phys. **30**, 277
3.129 Phillips, W.R. (1993) Nature 366, 4 November, p. 13

Information about nuclear interactions as derived from the theoretical and experimental study of nuclear level distributions, and symmetries governing the nuclear many-body system are discussed in the popular presentation

3.130 Weidenmüller, H. (1986) Comm. Nucl. Part. Phys. **16**, No. 4, 199

A detailed study of the derivation of level distributions as a function of matrix element distributions is given in the Vol. 1 of Bohr and Mottelson. We quote a couple of works that give access to the extensive literature on both experimental and theoretical level distributions.

On quantum chaos in a more general context:

3.131 Haake, F. (1991) Quantum Signatures of Chaos (Springer, Berlin Heidelberg)
3.132 Harney, H.L., Dittes, F.-M., Müller, A. (1992) Ann. Phys. (N.Y.) **220**, 159
3.133 Sieber, M., Steiner, F. (1991) Phys. Rev. Lett. **67**, 194

On the statistical properties of level distributions in atomic nuclei:

3.134 Bohigas, O. (1991) in *Chaos and Quantum Physics*, ed. by Giannoni, M., Voros, A., Zinn-Justin, J. (North-Holland, New York)
3.135 Bohigas, O., Weidenmüller, H. (1988) Ann. Rev. Nucl. Part. Sci., 421
3.136 Brody, T.A., Flores, J., French, J.B., Mello, P.A., Pandey, A., Wong, S.S.M. (1981) Rev. Mod. Phys. **53**, 385
3.137 Guhr, T., Weidenmüller, H. (1990) Ann. Phys. (N.Y.) **199**, 412
3.138 Haq, R.U., Pandey, A., Bohigas, O. (1982) Phys. Rev. Lett. **48**, 1086
3.139 Raman, S. et al. (1991) Phys. Rev. **C43**, 521
3.140 Shriner Jr., J.F., Bilpuch, E.G., Endt, P.M., Mitchell, G.E. (1990) Z. Phys. **A335**, 393
3.141 Weidenmüller, H. (1985) in *Nuclear Structure*, ed. by Broglia, R.A., Hagemann, G., Herskind, B. (North-Holland, Amsterdam)

Statistical properties derived from shell model and collective model studies:

3.142 Alhassid, Y., Whelan, N. (1991) Phys. Rev. **C43**, 2637
3.143 Alhassid, Y., Novoselsky, A. (1992) Phys. Rev. **C45**, 1677
3.144 Drödz, S., Speth, J. (1991) Phys. Rev. Lett. **67**, 529
3.145 Frazier, N., Brown, B.A., Zelevinsky, V. (1996) Phys. Rev. **C54**, 1665
3.146 Kusnezov, D., Brown, B.A., Zelevinsky, V. (1996) Phys. Lett. **B385**, 5
3.147 Rekstad, J., Tveter, T.S., Guttormsen, M. (1988) Phys. Rev. Lett. **65**, 2122
3.148 Whelan, N., Alhassid, Y. (1993) Nucl. Phys. **A586**, 42
3.149 Zhang, W.M., Feng, D.H. (1991) Phys. Rev. **C43**, 1127

The field of very hot nuclei also contains many facets; too many to go into detail here. We have singled out the dramatic case of the discovery of two-phonon excitations that "survive" at very high excitation energy inside the nucleus.

3.150 Auerbach, N. (1987) in Proc. Workshop on Pion–Nucleus Physics: Future Directions and New Facilities at LAMPF (American Institute of Physics) 163, 34
3.151 Auerbach, N. (1990) Ann. Phys. (N.Y.) **197**, 376
3.152 Boretzky, K. et al. (1996) Phys. Lett. **B384**, 30
3.153 Mordechai, S., Moore, S.F. (1991) Nature **352**, 293
3.154 Mordechai, S., Moore, S.F. (1994) Int. J. Mod. Phys. **E3**, 39

In reaching regions far from stability, besides the technical literature, a number of more popular texts have appeared recently:

3.155 Armbruster, P., Munzenberg, G. (1989) Scientific American, May, 36
3.156 Austin, S.M., Bertsch, G.F. (1995) Scientific American, June, 62
3.157 CERN Courier, (1994) March, 3
3.158 CERN Courier, (1995) December, 2
3.159 Hansen, P.G. (1993) Nature 361, February, 501
3.160 Hansen, P.G. (1993) New Scientist, 9th October, 38
3.161 Nitschke, M. (1989) New Scientist, 25th February, 55

Technical review papers and some original papers on light, exotic nuclei:

3.162 Hamilton, J.H. (1989) *Treatise on Heavy-Ion studies*, ed. by Bromley, D.A. (Plenum, New-York) Vol. 8, 1
3.163 Hansen, P.G. (1993) Nucl. Phys. **A553**, 89c
3.164 Hansen, P.G., Jonson, B. (1987) Europhys. Lett. **4**, 409
3.165 Hansen, P.G., Jensen, A.S., Jonson, B. (1995) Ann. Rev. Nuc. Sci. **45**, 591
3.166 Jonson, B. (1995) Nucl. Phys. **A583**, 733
3.167 Riisager, K. (1994) Rev. Mod. Phys. **66**, 1105
3.168 Tanihata, I. (1996) J. Phys. G **22**, 157

Recently, the doubly magic nucleus ^{100}Sn has been discovered; some of the keynote papers are:

3.169 Chartier, M. et al. (1996) Phys. Rev. Lett. **77**, 2400
3.170 Lewitowicz, M. et al. (1994) Phys. Lett. **B322**, 20
3.171 Schneider, R. et al. (1994) Z. Phys. **A348**, 241

Exploratory theoretical studies on how the nuclear mean-field might change as one approaches the drip-line are reported in the following recent papers:

3.172 Dobaczewski, J., Hamamoto, I., Nazarewicz, W., Seikh, J.A. (1994) Phys. Rev. Lett. **72**, 981
3.173 Dobaczewski, J., Nazarewicz, W., Werner, T.R., Berger, J.F., Chinn, C.R., Dechargé, J. (1996) Phys. Rev. **C53**, 2809
3.174 Nazarewicz, W., Werner, T.R., Dobaczewksi, J. (1994) Phys. Rev. **C50**, 2860
3.175 Otsuka, T., Fukunishi, N. (1996) Phys. Rep. **264**, 297

Great efforts continue to produce in the laboratory the heaviest nuclei possible in our universe:

3.176 Armbruster, P. (1985) Ann. Rev. Nucl. Part. Sci. **35**, 135
3.177 Münzenber, G. (1988) Rep. Prog. Phys. **51**, 57
3.178 Schmidt, K.-H., Morawek, W. (1991) Rep. Prog. Phys. **54**, 949

4. Nuclear Physics
with Electromagnetic Interactions

4.1 Layers of Constituents

As we have already discussed in Chaps. 2 and 3, a great deal of information about how protons and neutrons move inside the nucleus can be understood simply in terms of protons and neutrons interacting via an effective force inside the nuclear medium. Because of the presence of many other nucleons, this effective force deviates strongly from the corresponding force acting between free nucleons. However, a more fundamental theory would treat all medium effects explicitly, implying that, besides the pure nucleonic degrees of freedom, the mesonic degrees of freedom also start playing an important role. One should even take into account the various excited modes of the nucleons themselves. Eventually, the underlying building blocks of all matter involving the quark and gluon degrees of freedom will emerge in a description of how nucleons are built and, in a later step maybe, of how those composite particles (protons, neutrons) interact in an A-body nucleonic system. For most purposes though, as is clearly seen in the first three chapters, only effective forces and effective nucleons enter the description of the atomic nucleus. If one now wishes to find out at which energy (or length scale) particular new characteristics signaling the presence of mesons, excited nucleonic states and, even deeper, quark and gluon degrees of freedom actually emerge, one will need the appropriate microscope and a high enough energy to do so.

Thus we return to the first chapter where it was outlined that various 'layers' can be distinguished in the atomic nucleus (Fig. 4.1). On the scale of 5–10 fm, one observes the global nuclear features like surface effects and liquid drop behavior arising from the long-range correlations in the nucleon–nucleon forces. At a scale of about 1 fm, one observes that nucleons are moving in a correlated way because of the specific short-range correlations present in the nucleon–nucleon forces. Nucleons no longer behave independently. Furthermore, this is the length scale at which meson effects start to enter the picture, as well as excited baryonic states (Δ-resonance). This all results in a much more complex picture of the nucleus compared to the original independent-particle description. On a still smaller scale (from 1 down to 0.2 fm), one passes yet another boundary at which QCD effects start to modify the picture. On this scale, one has actually made the transition from regular nuclear physics to the higher energy physics scale where new phe-

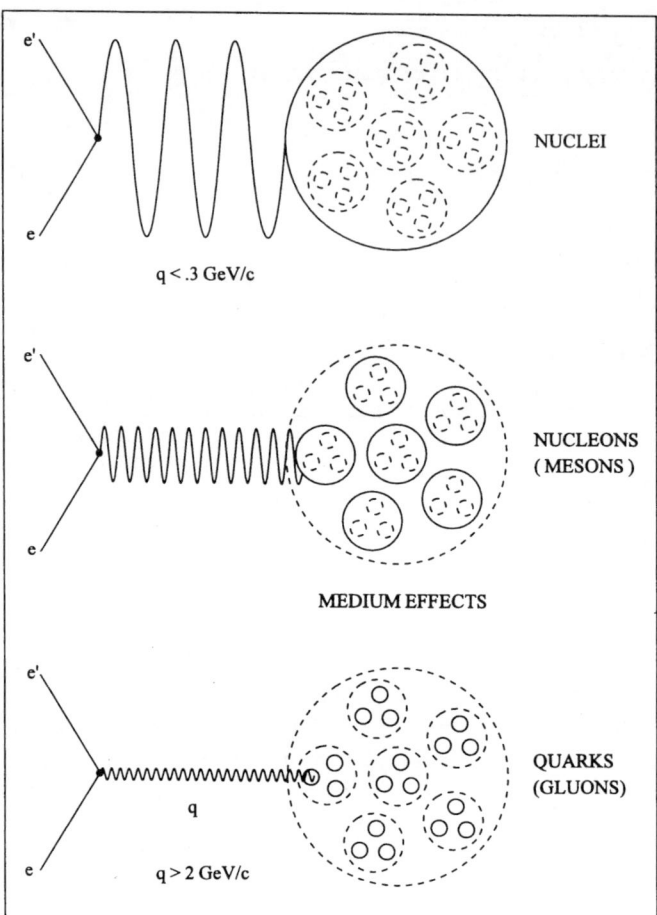

Fig. 4.1. Illustration of the use of the electromagnetic probe, as encountered in inelastic electron scattering (e, e'), to map out various scales inside the atomic nucleus. Using q as the variable characterizing momentum transfer ($\lambda \simeq 1.25/q$) with q in units GeV/c and λ in units fm. At the largest scale one can map large-scale and global properties ($q < 0.3$ GeV/c). At q values around 1 GeV/c, the specific nucleon properties and medium effects become visible. At the other extreme ($q > 2$ GeV/c), quark effects may well become observable. (Reprinted from A. Richter (1993) Nucl. Phys. **A553**, 417c. Elsevier Science, NL, with kind permission)

nomena are expected. One can even make this more quantitative: For highly relativistic electrons the spatial resolution is inversely proportional to the momentum transfer and is given by the relation

$$\lambda \simeq 1.25/q \ , \tag{4.1}$$

with q the momentum transfer in GeV/c.

At present, the length scale of 1–0.5 fm can be covered by electron facilities in Amsterdam (AMPS) and Mainz (MAMI) and at MIT (BATES) and the shorter scale can be covered by the recently commissioned CEBAF facility (4 GeV continuous wave electron superconducting accelerator), SLAC at Stanford, and HERA in Hamburg.

4.2 The Electromagnetic Probe

Photons are the force carriers of the electromagnetic interaction (Chap. 1) and they allow us to probe the nucleus in two different ways: by absorption of real photons or by the exchange of virtual photons in the scattering of charged particles off the nucleus. In the latter case, because of the use of virtual photons, the energy and momentum transfer to the atomic nucleus in a scattering event can be varied independently by changing the kinematical conditions. The electromagnetic force can act on the scale of the whole nucleus but also at the smaller scale of nucleons and so this interaction is useful for probing both global and local properties of the atomic nucleus. The scattering process can be visualized as in Fig. 4.2.

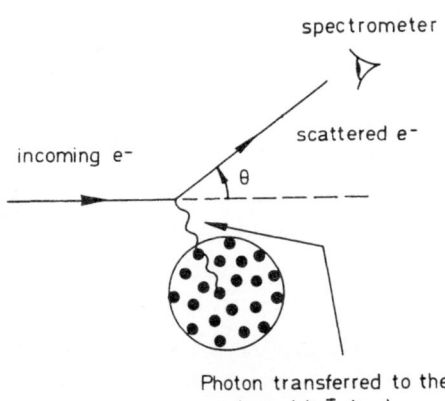

Fig. 4.2. Schematic picture of the scattering of an electron. In the electromagnetic interaction, energy $\hbar\omega$ and momentum $\hbar\mathbf{q}$ are exchanged with the nucleus through the exchange of a virtual photon. (Taken from K. Heyde *Basic Concepts in Nuclear Physic* ©1994 IOP Publishing, with permission)

The scattering process leads to a specific response of the nucleus. This response is a function of the energy and momentum transfer and represents the dynamical reaction of the composite system that is the atomic nucleus. One can distinguish three basic regions (Fig. 4.3). In the lower energy region, scattering on the ground state (elastic process) and on the various bound states (inelastic processes) can proceed and will be observed as a number of specific resonant structures in the full cross-section. Increasing the energy and adjusting the momentum transfer, a region is obtained near to 100 MeV in which the incoming virtual photon resonates on the 'quasi'-independent motion of a single nucleon that can be knocked out of its particular nuclear orbital. This region is called the "quasi-elastic" regime and is most interesting since it facilitates a detailed study of how the nucleons (both loosely and more strongly bound ones) move and behave inside the nuclear medium. We shall return to this point in Sect. 4.3. At the higher energies then, the first striking event occurs when the energy of the virtual photon exchanged is high enough to momentarily excite a nucleon into the excited delta-resonant state (Δ-resonance), which can subsequently decay by emitting a nucleon and a pion. This process of mapping out the response is quite general and is common to many other domains of physics when investigating how a given system reacts to external fields (here the electromagnetic field). By scanning through all the various domains and resonances, a unique view of the atomic nucleus – both of its global behavior and of the way in which nucleons are organized inside the nucleus – is obtained.

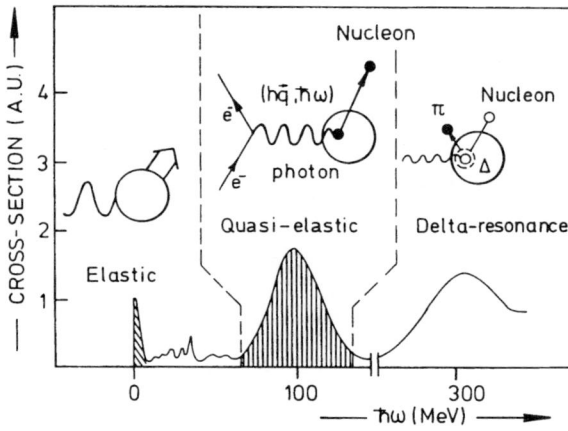

Fig. 4.3. The cross-section for electron scattering off the atomic nucleus as a function of the energy transferred ($\hbar\omega$ in units of MeV). A division into three major regions is presented, characterized by elastic, quasi-elastic, and Δ-resonance scattering. (Taken from P.K.A. de Witt-Huberts, C. De Vries *FOM Annual Report* ©1985, by courtesy of P.K.A. de Witt-Huberts)

Just recently, the Continuous Electron Beam Accelerator Facility, better known as CEBAF , went on-line. It represents one of the most powerful microscopes for looking deep inside the nucleus and for bridging the gap between the hadronic and quark descriptions of nuclear matter. We give some more technical details both about the facility as well as about the physics program at CEBAF in the following technical box (Box VIII).

Box VIII

CEBAF On-Line for Physics

After a long period of design, construction and commissioning, the Continuous Electron Beam Accelerator Facility (CEBAF) at Newport News, Virginia, is now on-line. The first experiments being set up will aim to bridge the gap between a description of nuclei using hadronic and mesonic degrees of freedom, on the one hand, and a quark description of nuclear matter, on the other hand.

We first give a few technical details. The continuous wave (CW) superconducting accelerator will operate at 4 GeV. A schematic outline of the facility is shown in Fig. VIII.1. The electrons are accelerated by successive passes through two 0.4 GeV antiparallel linacs that are linked by five recirculation arcs. A high current of 200 μA is expected. Over the whole complex about 2200 magnets are needed for the full accelerator and beam transport system. Because of the superconductivity requirements at CEBAF , it was necessary to build the world's largest 2K refrigerator source of superfluid helium. Finally the electron beam can be directed to one of the three experimental halls.

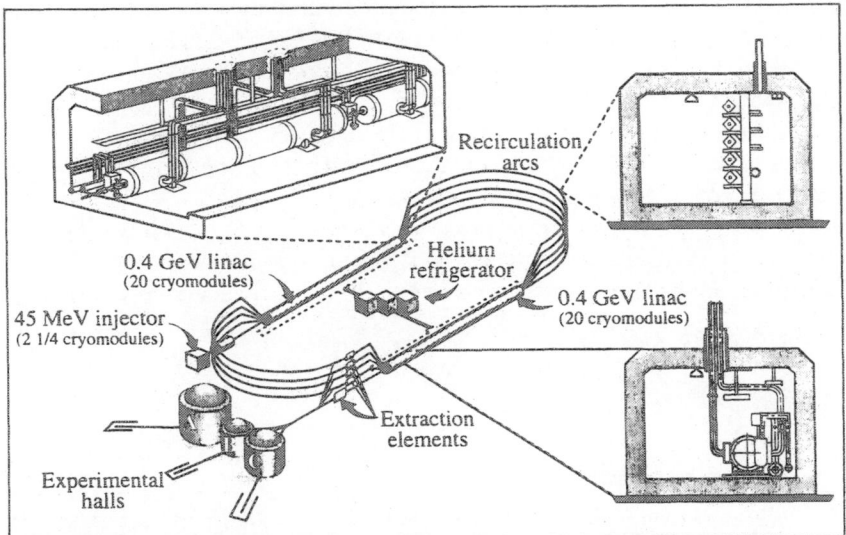

Fig. VIII.1. Schematic outline of the Continuous Electron Beam Accelerator Facility (CEBAF) superconducting accelerator, recently renamed the Thomas Jefferson National Accelerator Facility (TJNAF). Electrons are accelerated up to 4 GeV by successive passes through two 0.4 GeV antiparallel linacs linked by recirculation arcs. (Taken from CERN Courier (1996) March, with permission)

Concerning its physics program, the new 'microscope' should be able to study properties ranging from those of a large nucleus down to phenomena on the scale of one-tenth of the nucleon diameter. Thus, two large classes of experiments will be performed: one type aiming at a deeper understanding of the nucleon itself on the quark level (see also Sect. 4.4) and the other type trying to understand the bound nucleus, this time starting from the quark level upwards.

On the smallest scale, electrons can scatter off quarks inside the nucleon and, via this scattering process, information can be extracted about the spatial and momentum distributions of these quarks. At this level, the electrons can be used to promote the proton and/or neutron into an excited state and then study the subsequent decay process in detail (spectroscopy of baryons). Even though information is available on some of these topics, the 4 GeV CW CEBAF facility, as a dedicated accelerator, should be able to shed unprecedented light on the nucleon and reveal some of its innermost secrets.

In the second step, once the deeper level has been thoroughly studied and mapped out further, experiments will embark on the ambitious program of linking the nuclear structure to the quark structure. Therefore experiments will be designed to study charge and magnetization distributions at the very small length scales inside the nucleus. Other experiments are being designed to study the situation, where in nucleon language, protons 'overlap' inside the nucleus and thus composite systems can be knocked out of the nucleus in a single electron scattering event. The long-standing question of whether the properties of a free proton differ from those of one moving inside the nucleus, and, if so, how strongly, may well find a solution thanks to the very high resolving power of the CEBAF electron 'microscope'.

In addition, a number of experiments will be set up to search for more exotic forms of quark–gluon organization, other than the standard proton and neutron construction inside nuclei on the lower energy scale. An outline of the CEBAF experimental program is shown in Fig. VIII.2 where the two fundamental questions concerning (i) quarks confinement and (ii) the nucleon–nucleon force and its short-range behavior are clearly presented.

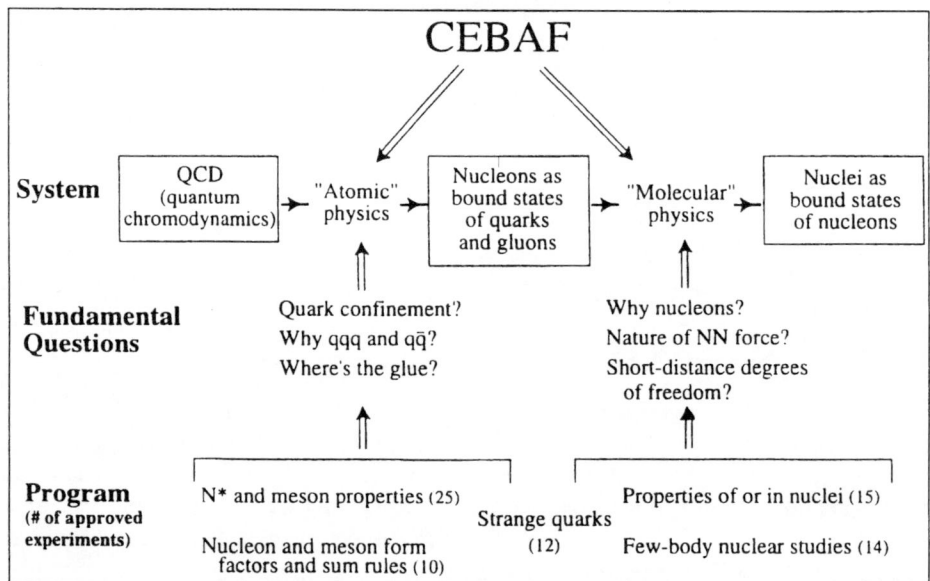

Fig. VIII.2. Outline of the experimental program to be carried out at the 4 GeV TJNAF. (Taken from CERN Courier (1996) March, with permission)

With the above unique characteristics of CEBAF, spin-offs and applications in various other domains of physics will quickly emerge. Even now, there is much interest in a high average power, tunable wavelength free electron laser (FEL). Here, both civilian and navy funding is contributing and manufacturers like DuPont, 3M, Xerox, and IBM are participating. Moreover, the recirculation arcs are such that, with appropriate modifications, energy upgrades to 8 GeV and even 16 GeV could well be realized. Before that the physics program expects to enjoy several years of new and probably very exciting experimental achievement.

4.3 How Nucleons Move Inside the Nucleus: Mean-Field Theory and Beyond

4.3.1 Independent Particle Motion and Nucleon Knock-Out

In the above sections, we have discussed the various levels in the atomic nucleus and thus also the energy regime in which the various nucleonic degrees of freedom can be optimally studied. In the quasi-elastic regime, where the energy and momentum of the incoming photon (real or virtual) are adjusted to the single-particle motion inside the nucleus, the process of scattering an electron off the nucleus and, at the same time, knocking a nucleon out of its orbit inside the nucleus, the very concept of independent particle motion (IPM) in an average one-body field can be tested critically. We depict this knock-out process in Fig. 4.4 where part (a) represents this IPM. At the same time we also illustrate the kinematics of the process in Fig. 4.5 where the one-photon exchange process of an electromagnetically induced one-nucleon knock-out is drawn and this in a plane-wave impulse approximation (PWIA).

In this latter most simple approach all nucleons, except one, are treated as spectators. Inside the nucleus the nucleon moves in a bound orbital corresponding to a given nuclear mean-field and then as an outgoing nucleon it moves in an undisturbed way thus described by a plane wave. Imposing momentum conservation in the two vertices and assuming the initial nucleus A to be at rest in the laboratory (thus with $\boldsymbol{p}_A = 0$), one has the relation

$$\boldsymbol{p}_m = \boldsymbol{p}_a - \hbar\boldsymbol{q} = -\boldsymbol{p}_B , \tag{4.2}$$

where $\boldsymbol{p}_a$ is the momentum of the outgoing nucleon and $\boldsymbol{p}_B$ the momentum of the residual nucleus. The momentum $\boldsymbol{p}_m$ is called the "missing" momentum and, in the quasi-free or independent particle picture, corresponds to the momentum of the outgoing nucleon just before it undergoes the interaction with the external electromagnetic field in which it absorbs the momentum and energy. By measuring this momentum distribution $\boldsymbol{p}_m$, one obtains information from which the nuclear wave function can be derived, both in momentum

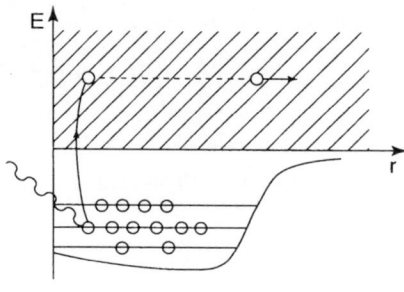

a

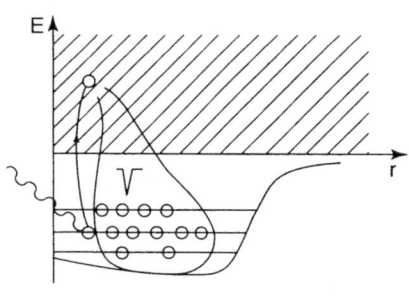

b

Fig. 4.4. Schematic illustration of a process in which a single nucleon is knocked out (one-nucleon knock-out) in an independent particle model (**a**), and in an improved description including Random Phase Approximation (RPA) correlations, the latter presented pictorially by the 'balloon' labeled with the nucleon–nucleon two-body interaction V (**b**). (Taken from J. Ryckebusch (1988) Thesis, Univ. of Gent, unpublished)

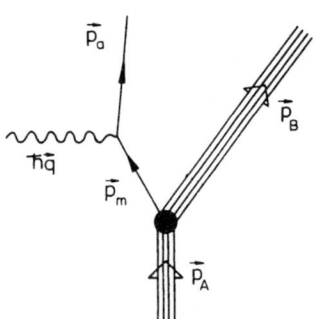

Fig. 4.5. Kinematics of a one-photon exchange ($\hbar q$) process in which a single nucleon is knocked out, with momentum p_a, leaving the residual nucleus in a state with momentum p_B (initial momentum of the nucleus is denoted by p_A). The process is described in the plane-wave impulse approximation (PWIA) and p_m (see text) denotes the 'missing' momentum, defined as $p_m \equiv p_a - \hbar q$. (Taken from K. Heyde *Basic Concepts in Nuclear Physic* ©1994 IOP Publishing, with permission)

and in coordinate space. This momentum distribution, which is measured by the spectrometers at a number of electron accelerator facilities like MAMI, AMPS, BATES, and now CEBAF, tells us about the velocity with which the nucleons move inside the nucleus. An example of such a distribution, technically called the 'spectral function' is shown in Fig. 4.6 for the case of proton knock-out from the very heavy nucleus ^{208}Pb. Inspecting these results, one can observe that, for nucleons moving with a speed up to about 1/4–1/3 of the speed of light, the momentum distribution is very well described in terms of an average field acting on all nucleons inside the nucleus. Deviations appear at the very end of the spectrum (the tail of the distribution). These deviations shed light on the various components in the nuclear force that

induce correlations (both long- and short-range correlations) that go beyond
the lowest-order mean-field description.

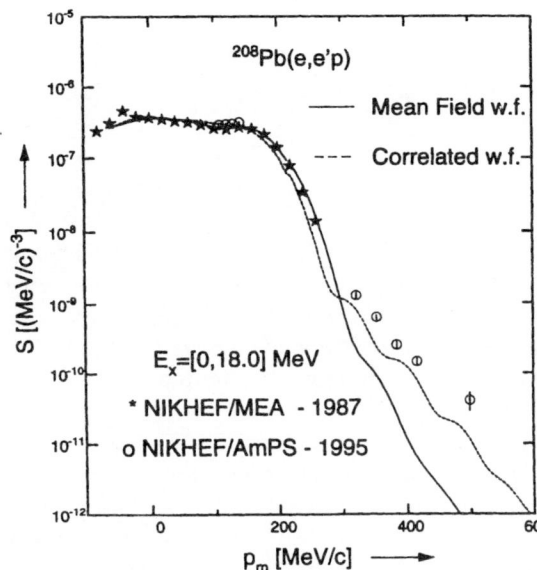

Fig. 4.6. The measured probability distribution S describing a proton moving with
momentum p_m in the nu-
cleus ^{208}Pb, together with the
results of theoretical calculations. The *full line* is the result
from a mean-field study; the
dashed line result includes
various types of nucleon–
nucleon correlations, in particular strong short-range correlations. (Reprinted from
L. Lapikas, G. van der Steen-
hoven *Nuclear Physics News*
Vol. 6, No. 2 ©1996 Gor-
don & Breach, N.Y., with
permission

As a short technical aside, we note that the momentum distribution can be
transformed, within the framework of the independent particle model into the
corresponding coordinate wave function for the nucleon. This process is called
Fourier–Bessel transformation and one can thereby relate a nucleon moving in
an independent particle orbital (characterized by quantum numbers n, l, j, m
and denoted by a for short) to its corresponding momentum distribution via
the transform

$$\rho_a(p) = \frac{1}{2\pi^2\hbar^3} \left[\int dr r^2 j_{l_a} (pr/\hbar) \, \varphi_{n_a l_a j_a} \right]^2 v_a^2 (2j_a + 1) \, , \qquad (4.3)$$

where $j_l(x)$ denotes the spherical Bessel function and v_a^2 the occupation probability in the nucleus of the orbital a. This is a very powerful method which
relates the fine details in the nuclear wave functions to certain components
in the momentum distribution and vice versa.

Because of the intuitive appeal of this quasi-free process, a lot of experimental effort has gone into detailed measurements over a large range of
momentum transfer, so spanning the region from very light nuclei up to the
heavy doubly-closed shell nucleus ^{208}Pb. The analysis has given a strong experimental underpinning of the nucleonic motion inside the nucleus, at least
for nucleons that are knocked out of weakly bound orbitals with an energy
close to the Fermi energy for a given nucleus. As was already becoming clear,

for the nucleons moving at a very high speed and/or in very deeply bound states, deviations from such an independent particle picture inevitably show up. We now discuss some ways of advancing beyond the lowest order mean-field picture.

4.3.2 Beyond the Mean Field: Nucleon–Nucleon Correlations and Deep-Hole States

The picture that was given in Fig. 4.4a actually needs to be extended (as in part b) because the remaining A nucleons are not just spectators. The nucleon–nucleon interaction (represented by the balloon and labeled with the letter V) will induce certain correlations in the motion of the nucleons inside the nucleus beyond the independent particle motion. It is these same correlations that give rise to coherent motion and thus to the microscopic cause of the nuclear collective modes like vibrations of various multipolarity, rotational motions, and the various coupled modes. So, the picture becomes considerably modified when these long-range correlations (technically called Random Phase Approximation RPA) are taken into account. We now explain the effect of those RPA correlations in an intuitive way: Initially, we start from the same point: a nucleon gets excited into a continuum state. The excited nucleon will now feel the other nucleons through the presence of an average field but also through direct and specific components of the nucleon–nucleon force. This residual interaction can be seen as a way of further exchange of energy and momentum between the excited nucleon and the remaining $A - 1$ nucleons on its way to being knocked out of the nucleus. Quite complicated and detailed processes can occur at this stage but the final picture leads to the concept of a 'doorway' state lasting long enough for a nucleon to be ejected with the final nucleus remaining in a definite quantum state. It is interesting to note that, by definition, the emitted nucleon will not be the same one that was originally excited from its motion in a nuclear orbital. So, the present picture goes well beyond the original IPM picture and includes a special class of nucleon–nucleon correlations called RPA correlations. A dramatic illustration of the modifications brought about by these RPA correlations is given in Fig. 4.7 where, for ^{16}O, we compare the knock-out processes of a proton and a neutron. Within an independent particle approach, the emission probabilities for a neutron and for a proton are very different; the RPA correlations, however, cause the two probability distributions to become much more similar, as is observed experimentally. Thus there is clear support for the inclusion of RPA correlations going beyond the mean-field approach. Put another way, the presence of the nuclear medium cannot simply be taken out as a spectator but influences a number of processes in a fundamental way. One influence is to make the knock-out probabilities for a proton and a neutron under similar kinematical conditions more equal.

A second source of deviations from the simple shell model picture of a nucleon knocked out of a single-particle orbital has to be considered. If a

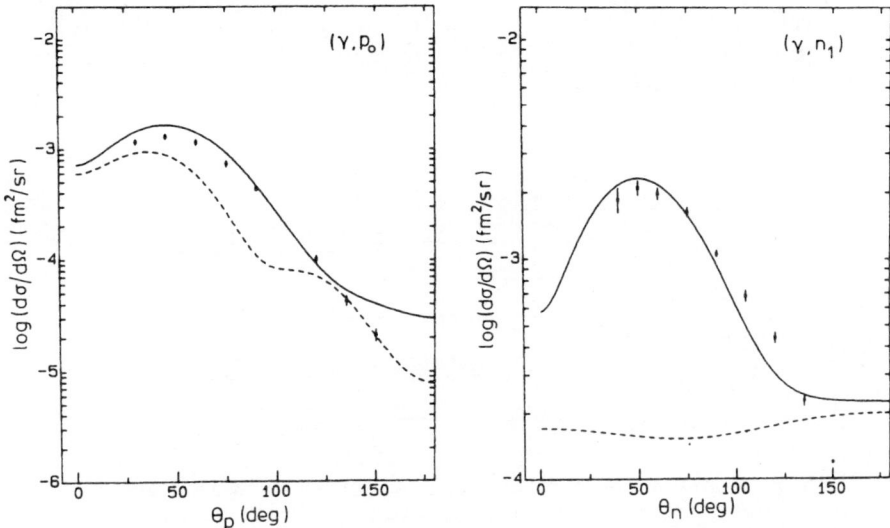

Fig. 4.7. Angular distributions describing the emission of a proton (γ, p_0) or of a neutron (γ, n_1) from ^{16}O, using incoming photons with an energy of 60 MeV. The *dashed lines* give the results of a Hartree–Fock (HF) mean-field calculation and strongly differ from one another. The *full lines* are the result of a combined HF+RPA calculation and are seen to be much more similar. The experimental data are also shown. (Reprinted from J. Ryckebusch et al. (1987) Phys. Lett. **194B**, 453. Elsevier Science, NL, with kind permission)

nucleon were knocked out of a specific single-particle state with a certain binding energy, the energy dependence of the knock-out probability should reduce to a Dirac delta-function at the corresponding binding energy. In practice, however, the single-particle strength function becomes spread out over a certain energy interval depending on the binding energy of the particular state and the mass of the nucleus. This process can be depicted schematically as follows (Fig. 4.8). At the energy of the single-particle state from which a nucleon is removed, many more complex nuclear configurations can exist, ranging in general over various many-particle–many-hole configurations. This means that, in the vicinity of the single-hole configuration, a rather high density of such complex configurations are obtained. The residual interaction will automatically smear out the original single-hole strength and dissolve it in a background of many complex states. Now, depending on the specific mass of the nucleus one is studying, and on the coupling strength, the original strength can be spread out in a variety of ways. A simple and yet well-defined way of including this residual coupling is through a coupling of the single-hole motion with nuclear surface vibrational excitations. The technical details are relevant not just to nuclear physics, but appear in a lot of quantum many-body systems where the elementary modes of motion are imbedded in a variety of more complex excitations. Thus the spreading of the elementary modes, in which they acquire a given energy spread and decay width (if unbound), is a very general phenomenon that occurs in solid state, atomic, and molecular physics.

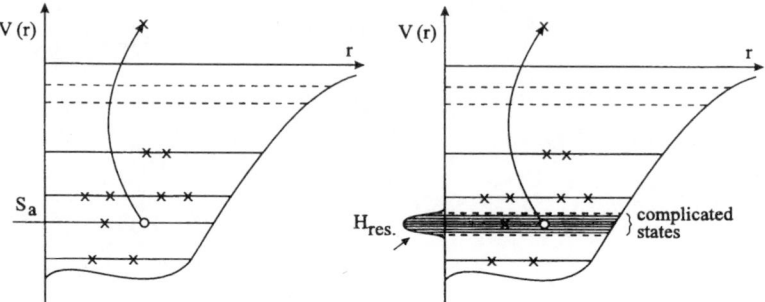

Fig. 4.8. Pictorial description of the process in which a nucleon is removed from its specific orbital a in the average one-body field (*left*). In the more realistic picture (*right*), a particle, moving in a single-particle state a, interacts with a large number of close-lying more complex configurations. (Taken from K. Heyde *The Nuclear Shell Model* ©1994 Springer, Berlin Heidelberg, with permission)

    To illustrate the above mechanism, we show the results of a micro-scopic calculation in which the fragmentation of single-hole motion over a background of increasingly more complex 2h-1p excitations is derived using Green's function techniques. In Fig. 4.9, the subsequent phases of damping the independent particle motion inside the nucleus over a reservoir of many complex excitations and the final convergence are well illustrated.

    This field of testing the limits of single-particle motion as a first but often valid approximation to the more complex interacting many-nucleon system is a most lively one. It also shows the power and potential of electromagnetic interactions for probing the ultimate limits of the mean-field description. At the same time, the more detailed results clearly tell us to go beyond the average field and include various types of correlations which improve the inital IPM description, and give a very firm support of the shell model description of nucleons moving inside the nucleus.

4.4 Physics at Even Higher Energies: Inside the Nucleon

The endeavor to understand hadrons is also important as an avenue to better appreciate nuclear physics on a fundamental level. It is clear that, in order to understand the nucleon–nucleon force, a deep understanding of the nucleon structure is a prerequisite. Even though nuclei are composed of nucleons and nucleons are built out of quarks, it is not clear at all whether and where one needs quark degrees of freedom in order to understand the behavior of

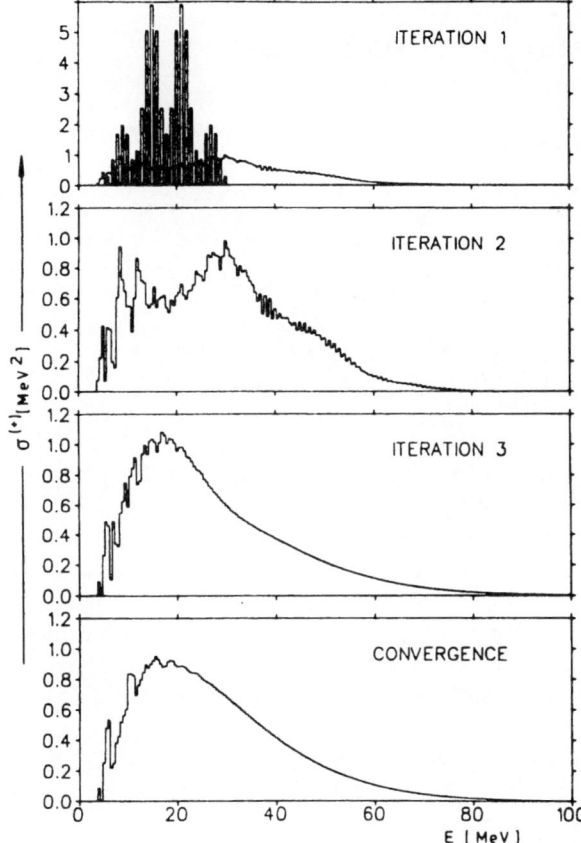

Fig. 4.9. Results from a complicated Green's function model calculation in which the 1-hole states couple to more complicated 2-hole–1-particle states and thus cause a spreading of the initial 1-hole strength. The figure indicates the subsequent steps (iterations 1,2,3 up to convergence) in deriving a stable 1-hole strength distribution. (Reprinted from D. Van Neck et al. (1991) Nucl. Phys. **A530**, 347. Elsevier Science, NL, with kind permission)

atomic nuclei. A large number of properties can be well understood in terms of nucleons and mesons only. Thus, to conclude that quark degrees are needed explicitly, dedicated and very well designed experiments must first be set up and carried out.

We list below a number of issues that are central in the investigation of hadron properties. We also highlight the specific technical elements that allow one to pinpoint quark properties inside the nucleon.

- Hadron spectroscopy
  - $B = 0$: meson spectroscopy, meson–meson interaction, multiquark and multigluon states, ...
  - $B = 1$: baryon spectroscopy, meson-baryon interaction, ...
  - $B = 2$: di-baryon and exotic baryon (H-particle) states, baryon–baryon interactions, ...
  - $B > 2$: multiquark configuration, nature of the confinement, ...

- Elementary properties of particle
  - Polarizabilities, electric and magnetic form factors, ...
- Nucleons and mesons in nuclei

- Wave functions and structure of hadronic systems
  - Medium effects
  - Color transparency
  - The Nucleon and its resonances
- The quark as a color probe
  - Propagation and hadronization in matter
- Facilities
  - PSI, LAMPF, TRIUMF, SATURNE, LEAR, CELSIUS, COSY, DAΦNE, MAMI, AMPS, BATES, CEBAF, SLAC, EEF,...

(List reprinted from A. Richter (1993) Nucl. Phys. **A553**, 417c, with kind permission from Elsevier Science, NL)

In the following sections we will discuss important inelastic processes: inelastic electron–proton scattering, exciting the proton itself, and studying, e.g., the proton's spin content. First, we briefly introduce some definitions.

In the scattering of a lepton ℓ ($\ell \equiv e, \mu, \nu, \bar{\nu}$, ...) off a nucleon, denoted by N, into a final lepton ℓ' and 'hadronic' state X, we define the following quantities (Fig. 4.10). Starting from $q \equiv k - k'$, the four-momentum of the exchanged particle, one derives the various quantities listed in Table 4.1. They are given in (i) a frame where the target is fixed (laboratory frame), and (ii) the invariant relativistic frame.

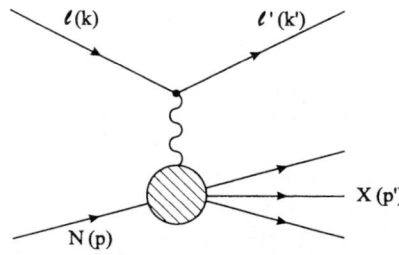

Fig. 4.10. Kinematics of a process where a lepton ℓ (with four-momentum k) scatters off a nucleon N (with four-momentum p) and emerges as a lepton ℓ' (with four-momentum k') giving rise to a final hadronic state X (with four-momentum p'). The various kinematical relations are given in Table 4.1

Table 4.1.

(i)	(ii)
$Q^2 \equiv 4EE' \sin^2 \theta/2$	$Q^2 = -q \cdot q$
$\nu \equiv E - E'$	$\nu = \frac{p \cdot q}{\sqrt{p \cdot p}}$
$x \equiv \frac{Q^2}{2m_{\mathrm{p}}\nu}$ (Bjorken x)	$x = -\frac{q \cdot q}{2p \cdot q}$
$y \equiv \frac{\nu}{E}$ (Bjorken y)	$y = \frac{p \cdot q}{p \cdot k}$
$\omega^2 = m_{\mathrm{p}}^2 + 2m_{\mathrm{p}}\nu - Q^2$	$\omega^2 = (p + q)^2$

4.4.1 Inelastic Electron–Proton Scattering

In the simple picture of inelastic electron–proton scattering a virtual photon is exchanged with the proton in the proton rest frame, exchanging energy and momentum. This virtual photon has a mass which obeys the relation $m_\gamma^2 = -Q^2$. If now the momentum transfer is so large that the corresponding length that can be resolved in the scattering event is small compared to the proton size ($Q \gg 1$ GeV), then the photon no longer interacts with the proton as a whole but with one of its quarks. In the simplest of situations, the photon scatters elastically off a quark which carries a fraction x of the proton momentum and is related to Q^2 and ν via the relation

$$x = \frac{Q^2}{2m_\mathrm{p}\nu} \, , \tag{4.4}$$

where m_p is the proton mass. The quark that is knocked out and the remaining part of the proton cannot be observed as 'real' particles (because they carry color charge) (Fig. 4.11). Both then give rise to jets of strongly interacting particles such as nucleons, pions, and kaons.

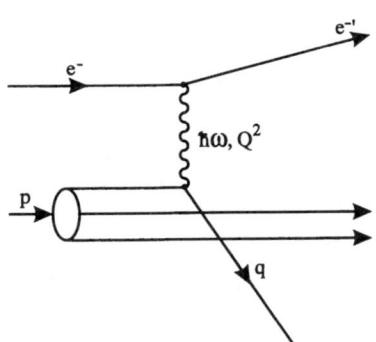

Fig. 4.11. Kinematics of electron–nucleon scattering in which energy ($\hbar\omega$) and momentum (described by $Q^2 \equiv 4EE' \sin^2(\theta/2) = -q \cdot q$) are transferred to a quark that gets knocked out leaving a remnant part of the nucleon. Neither the quark nor the remnant nucleon can be observed, but both give rise to jets of strongly interacting particles

Information on the density distribution of quarks inside the proton $q(x, Q^2)$ can be obtained from the structure function of the proton

$$F(x, Q^2) = \sum_q e_q^2 x q(x, Q^2) \, , \tag{4.5}$$

which is directly measured in deep inelastic electron scattering . Here e is the electric charge of the quark q (2/3 for the up quark and $-1/3$ for the down quark). The measured distribution clearly depends on the momentum transfer: as this momentum transfer increases, more and more fine details inside the proton can be observed.

The first major achievement of these electron scattering experiments was the discovery at SLAC in 1969 that the proton is actually made out of point-like constituents which were later identified as quarks and gluons. The electron–proton collider HERA which has been in operation since 1992 at DESY (Hamburg) has improved the resolving power in electron–proton scattering by more than one order of magnitude compared with the former SLAC experiments. Here, 30 GeV electrons collide head-on with 820 MeV protons.

4.4.2 Excitation Spectrum of the Proton

The above electron–proton scattering experiments have led to the picture of a proton as consisting of three constituent quarks orbiting around each other (Fig. 4.12). One has to remember that those constituent quarks are effectively "dressed" quarks and so are like quasi-particles in a general many-body description. Here the original quark becomes dressed with quark–antiquark pairs and additional gluons moving around.

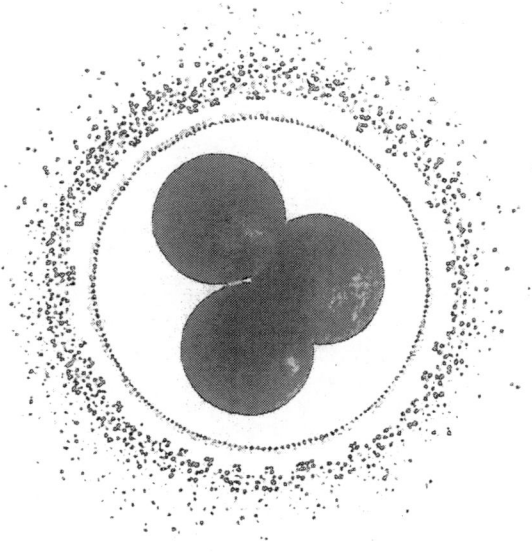

Fig. 4.12. The nucleon. This artist's impression shows three quarks in a spherical "bag" surrounded by a cloud of mesons. The quarks themselves should be thought of as point-like particles; their finite extent in the picture may be taken as an indication of their probability distributions. (Adapted from G.E. Brown, M. Rho (1983) *Physics Today* February 24)

In further testing this picture, one is interested in exciting the proton or neutron into its various excited states like the Δ (1232 MeV), N (1440 MeV), or N (1520 MeV) configurations. In these excitations one or more of the quark spins is flipped. Calculations at this level predict various non-spherical shapes and, if this is indeed the case, one might explore polarizability of this specific excited nucleon configuration. We illustrate various model forms for the excitation spectrum of the proton as well as the total photoabsorption cross section of the proton in Fig. 4.13.

4.4.3 Proton Spin

Protons and neutrons are fermions characterized by intrinsic spin 1/2 (in units $\hbar$) and this spin is also the basis for understanding the spin of more complex bound systems of protons and neutrons, i.e., atomic nuclei. Early attempts to understand the nucleon spin considered it as the sum of the spin 1/2 of each of the three constituent quarks that make up the nucleon. This yielded correct results both for the proton and neutron spin and for the magnetic moment. We have to keep in mind though, that on a more fundamental level, even the proton and neutron have to be considered as highly complex many-body interacting systems of quarks, antiquarks, and gluons and that, in terms of these more elementary constituents, it might be more difficult to understand the nucleon spin (see Fig. 4.14)

Recent polarization experiments using deep inelastic electron and muon scattering off a nucleon all indicate that the spin carried by the quarks seems to be less than 1/3 of the nucleon spin and much less than suggested to be the case in a constituent quark model. One does have a sum rule, however: The total intrinsic spin of all the internal constituents added to the orbital motion angular momentum should give exactly spin 1/2! There seems to be no doubt at present about the validity of the underlying QCD theoretical description which has been borne out by tests of Bjorken's sum rule.

The very small amount of spin carried by the quarks and antiquarks taken together has motivated further detailed studies to probe the origin of this "missing spin" or of the "spin crisis", as it is often called. The aim in these experiments will be to single out the various spin contributions of different types of quarks and antiquarks, of gluons, and of orbital motion. The HERMES experiment at the HERA electron accelerator in Hamburg will use polarized nuclei such as ^{3}He with the aim of identifying different types of quarks. Experiments with high energy proton beams at the new RHIC accelerator (see also Chap. 5) aim to unravel the gluon spin content. Other experiments planned at MIT and CEBAF will seek a quantitative understanding of the proton's intrinsic magnetic properties associated with the strangeness quark content.

All these experiments are looking for the various angular momentum contributions that make up the spin 1/2 of the proton and the neutron. Let us recall the fact that the internal structure of the individual proton and neutron is as complicated a many-body problem as that presented by the atomic nucleus itself, where one is dealing with nucleon and meson degrees of freedom only. Future years are expected to bring many detailed experimental results from the various electron accelerator facilities.

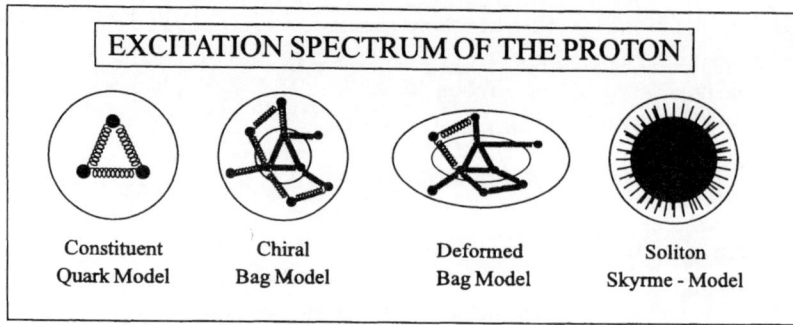

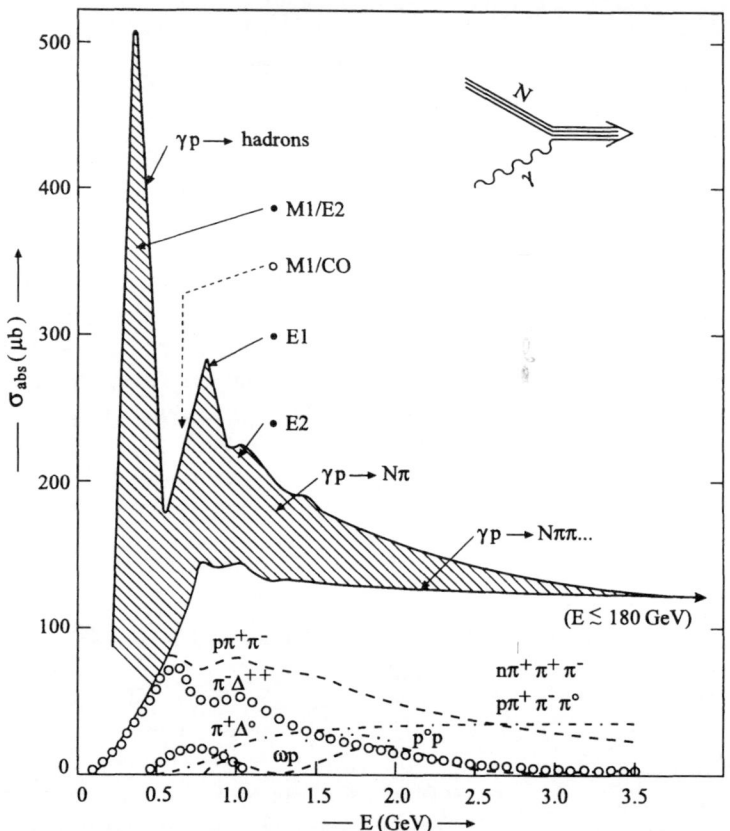

Fig. 4.13. The *upper part* shows schematically various quark confining configurations that may represent a single proton. The *lower part* illustrates the total photoabsorption cross-section (in units of microbarns, μb) of a proton as well as for a number of subprocesses. (Reprinted from A. Richter (1993) Nucl. Phys. **A553**, 417c. Elsevier Science, NL, with kind permission)

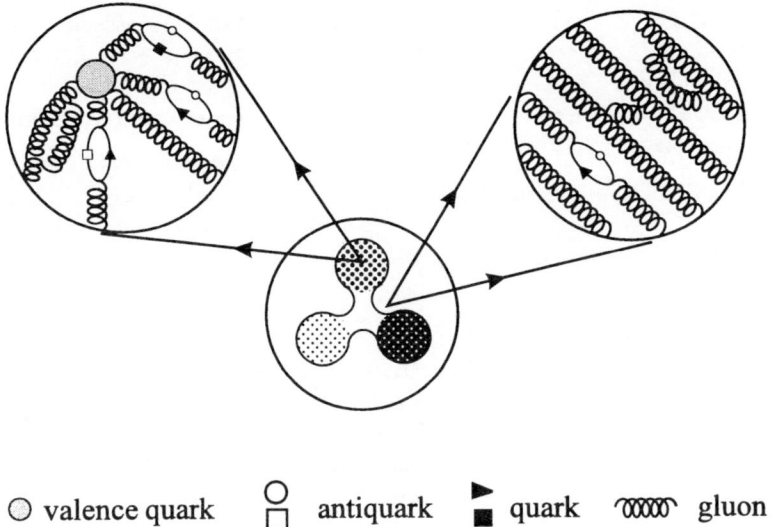

○ valence quark   ⊖ antiquark   ▶ quark   ⟋⟍⟍⟍⟍ gluon

Fig. 4.14. Schematic nucleon structure. The central nucleon is shown as a collection of three constituent quarks interacting via the color force. Closer views show a more complex picture. To the *left*, the single quark is accompanied by processes in which gluons are emitted and absorbed, or, where gluons materialize from quark–antiquark pairs. The *right-hand part* shows the gluons that carry the color force and themselves give rise to quark–antiquark pairs. (Adapted from B. Frois et al. (1994) *Physics World* July. IOP Publishing, with kind permission)

4.5 Further Reading

Electromagnetic interactions with real photons or using virtual photon exchange between scattering charged particles (particularly electrons scattering off nuclei) are described in a number of general texts and review papers:

4.1 Arenhövel, H., Drechsel, D. (eds.) (1979) *Nuclear Physics with Electromagnetic Interactions* (Springer, New York)

4.2 Cannata, F., Überall, H. (1980) *Giant Resonance Phenomena in Intermediate-Energy Nuclear Reactions*, in: Springer Tracts in Modern Physics, Vol. 89 (Springer, Berlin Heidelberg)

4.3 Donnelly, T.W., Walecka, J.D. (1975) Ann. Rev. Nucl. Sci. **25**, 329

4.4 Donnelly, T.W. (1996) Adv. Nucl. Phys. **22**, 37

4.5 Drechsel, D., Giannini, M.M. (1989) Rep. Prog. Phys. **52**, 1083

4.6 Eisenberg, J.M., Greiner, W. (1970) *Excitation Mechanisms of the Nucleus* (North-Holland, Amsterdam)

4.7 de Forest Jr., T., Walecka, J.D. (1966) Adv. Phys. **15**, 1

4.8 Faessler, A. (ed.) (1995) *Electromagnetic Probes and the Structure of Hadrons and Nuclei*, Progr. Part. Nucl. Phys. **34**

4.9 Fuller, E.G., Hayward, E. (1976) *Photonuclear Reactions* (Dowden, Hutchinson, & Ross, Stroudsburg, PA)

4.10 Hofstadter, R. (ed.) (1963) *Electron Scattering and Nuclear and Nucleon Structure, a Collection of Reprints with an Introduction*, (Benjamin, New York)
4.11 Überall, H. (1971) *Electron scattering from Complex Nuclei* (Academic Press, London)

Over the years electron scattering off nuclei on the scale of the entire nucleus and concentrating on nucleon motion, has given rise to a large body of information on nuclear structure properties. We list a number of review papers in this domain:

4.12 Benhar, O., Pandharipande, V.R. (1993) Rev. Mod. Phys. **65**, 817
4.13 Cavedon, J.M. et al. (1987) Phys. Rev. Lett. **58**, 195
4.14 Frois, B., Papanicolas, C.N. (1987) Ann. Rev. Nucl. Part. Sci. **37**, 133
4.15 Heisenberg, J. (1981) Adv. Nucl. Phys. **12**, 61
4.16 Heisenberg, J. (1984) Comments Nucl. Part. Phys. **13**, 267
4.17 Heisenberg, J., Blok, H.P. (1983) Ann. Rev. Nucl. Part. Sci. **33**, 569

Recently, CEBAF, a 4-GeV accelerator in Newport-News (Virginia, USA) has become operational and will push the limit of energy and length scale using CW facilities even further:

4.18 CEBAF (1986) Briefing for the Georgia Institute of Technology, April
4.19 Cern courier (1996) CEBAF on line for physics, March 12
4.20 Jacob, M. (1991) *Hadron Physics with electrons beyond 10 GeV*, Nucl. Phys. **A532**, 45c
4.21 Walecka, J.D. (1987) Summer workshop. CEBAF – A laboratory for Nuclear Physics

Electromagnetic interactions with nucleon (or multi-nucleon) knock-out have provided a unique way of testing nucleon motion inside the nuclear average field. Tests of the independent-particle model (IPM), of correlations beyond this and experiments to detect nucleonic excitations and mesonic components in the exchange currents have been carried out. As references we give a recent textbook and some of the major research papers.

4.22 Boffi, S., Giusti, C., Pacati, F.D., Radici, M. (1996) *Electromagnetic Response of Atomic Nuclei*, Oxford Studies in Nuclear Physics, Vol.20 (Clarendon Press, Oxford)

4.23 Boeglin, W.U. (1995) in *Hadrons in Nuclear Matter*, Proc. of the Int. Workshop XXIII on Gross Properties of Nuclei and Nuclear Excitations ed. by Feldmeier, H., Nörenberg, W. (GSI) p. 78
4.24 Boffi, S., Giusti, C., Pacati, F.D. (1993) Phys. Rep. **226**, 1
4.25 Dieperink, A.E.L., de Witt Huberts, P.K.A. (1990) Ann. Rev. Nucl. Part. Sci. **40**, 239
4.26 Gari, M., Hebach, H. (1981) Phys. Rep. **72**, 1
4.27 Mahaux, C., Sartor, R. (1990) Adv. Nucl. Phys. **239**, 57

4.28 Ryckebusch, J., Waroquier, M., Heyde, K., Ryckbosch, D. (1987) Phys. Lett. **B194**, 453

4.29 Ryckebusch, J., Waroquier, M., Heyde, K., Moreau, J., Ryckbosch, D. (1988) Nucl. Phys. **A476**, 237

4.30 Saruis, A.M. (1993) Phys. Rep. **235**, 57

4.31 Van Neck, D., Waroquier, M., Ryckebusch, J. (1990) Phys. Lett. **B249**, 157

4.32 Van Neck, D., Waroquier, M., Ryckebusch, J. (1991) Nucl. Phys. **A530**, 347

4.33 de Witt Huberts, P.K.A. (1990) J. Phys. **G16**, 507

4.34 A popular account is given in the NIKHEF documentation map (1996) to be obtained at NIKHEF, Postbus 41882, 1009 DB Amsterdam or at World Wide Web http://www.nikhef.nl

Electron scattering at high energies and in the deep inelastic regime with the aim not only of studying the nucleus but also of accessing interior properties of the nucleon itself has given rise to both some Nobel prizes in Physics and to a wealth of most exciting phenomena.

First the Nobel prizes. Descriptions of the work involved can be found in:

4.35 Hofstadter, R., Fechter, H.R., McIntyre, J.A. (1953) Phys. Rev. **91**, 422

4.36 Hofstadter, R., McAllister, R.W. (1955) Phys. Rev. **98**, 217

4.37 Friedman, J.I. (1991) Rev. Mod. Phys. **63**, 615

4.38 Kendall, H.W. (1991) Rev. Mod. Phys. **63**, 597

4.39 Taylor, R.E. (1991) Rev. Mod. Phys. **63**, 573

Some popular articles on proton structure studied with electron scattering are:

4.40 Brown, G.E., Rho, M. (1983) Physics Today, February, p. 24

4.41 Eisele, F., Wolf, G. (1994) Beam Line Vol. 24, No. 3, 29

4.42 Frois, B., Karliner, M. (1994) Physics World, July, p. 44

4.43 Gasiorowicz, S., Rosner, J.L. (1981) Am. J. Phys. **49**, 954

4.44 Gerard, A. (1990) Clefs CEA 16, 27

4.45 Martin, A.D. (1995) Contemp. Physics, Vol. 36, No. 5, 335

A few books, concentrating on properties of the nucleon itself have already been given in Chap. 2. Some additional books dealing with mesons in nuclei are:

4.46 Ericson, T.E.O., Weise, W. (1988) *Pions and Nuclei* (Clarendon Press, Oxford)

4.47 Rho, M., Wilkinson, D. (1979) *Mesons in Nuclei* (North-Holland, Amsterdam)

We also include a number of basic articles describing experimental techniques and the theory of deep inelastic electron scattering.

First some articles from Comments on Nuclear and Particle Physics reviews

4.48 Frois, B., Mathiot, J.F. (1989) Comments Nucl. Part. Phys. **18**, 307
4.49 Holstein, B. (1992) Comments Nucl. Part. Phys. **20**, 301
4.50 Jaffe, R.L. (1984) Comments Nucl. Part. Phys. **13**, 39
4.51 Leader, E. (1995) Comments Nucl. Part. Phys. **21**, 323
4.52 Sick, I. (1988) Comments Nucl. Part. Phys. **18**, 109

An interesting set of lecture notes is

4.53 Varvell, K. (1995) Lepton–Nuclear Scattering, Nuclear and Particle Physics (NUPP) Summer School, Melbourne, Lecture Notes (unpubl.)

Some basic articles on Bjorken scaling, the European Muon Collaboration (EMC), SMC, and on the spin structure of the proton:

4.54 Adeva, B. et al. (1994) Phys. Lett. **B320**, 400
4.55 Ashman et al. (1988) Phys. Lett. **B206**, 364
4.56 Ashman et al. (1990) Nucl. Phys. **B328**, 1
4.57 Bjorken, J.D.(1966) Phys. Rev. **148**, 1467
4.58 Bjorken, J.D. (1970) Phys. Rev. **D1**, 1376
4.59 Ellis, J., Jaffe, R.L. (1974) Phys. Rev. **D9**, 1444
4.60 SMC (1994) Nucl. Instr. Meth. **A343**, 363

The remaining citations are some review papers on the domain of intermediate and higher energy electromagnetic interactions concentrating on the scale of the nucleon and on the mesonic degrees of freedom

4.61 Christillin, P. (1990) Phys. Rep. **190**, 63
4.62 Day, D.B., McCarthy, J.S., Donnelly, J.W., Sick, I. (1990) Ann. Rev. Nucl. Part. Sci. **40**, 357
4.63 ELFE Project (1993) eds. Arvieux, J., De Sanctis, E. (Italian Physical Society, Bologna)
4.64 Giannini, M.M. (1991) Rep. Progr. Phys. **54**, 453
4.65 Krewald, S., Nakayama, K., Speth, J. (1988) Phys. Rep. **161**, 103
4.66 Mulders, P.J. (1990) Phys. Rep. **185**, 83
4.67 Spin stucture in high-energy processes (1994) Proc. of the 21th Summer Inst. Part. Phys. (SLAC Report-444, CONF-938767, UC-414 (T/E))

5. Exploring Nuclear Matter at High Densities

5.1 Introduction

As was discussed in Chap. 4, it is possible to study the fine details of the nuclear many-body system using electromagnetic probes. One can go beyond this and even observe the mesonic degrees of freedom explicitly, or even deeper to study quark degrees of freedom in some detail. These observations, however, are restricted to very local regions inside the atomic nucleus.

A totally different method for investigating the various characteristics inside the nucleus involves creating a hot, compressed, and highly excited form of nuclear matter (Fig. 5.1). This heating can act on the whole nuclear volume and thereby create possible new forms of nuclear matter. Access to such extreme states is provided by colliding heavy ions, as can be done at a number of accelerators (e.g., AGS at Brookhaven and SPS at CERN). The hope is to study such collisions at more dedicated installations like RHIC (the Relativistic Heavy Ion Collider at BNL) and, further in the future, at the LHC at CERN.

A number of basic properties of nuclear matter such as the *central density* were experimentally established during the 1950s and 1960s via electron scattering. The result for the central density was 0.17 ± 0.01 nucleons/fm^3. Information on the *compressibility*, another very important parameter characterizing nuclear matter, at the saturation density in atomic nuclei, has been deduced more recently from studying the compression modes of the nucleus, e.g., the monopole breathing mode. A value for the compression modulus K of 200–300 MeV for bulk nuclear matter was thus derived. A third basic parameter is the *binding energy per nucleon in nuclear matter* with a value of 16 ± 0.5 MeV.

The various properties of the nuclear many-body system like density, pressure, p, and internal temperature, T, are not independent variables like in any other thermodynamic system. They are related via an equation of state (EOS). A complication in describing atomic nuclei derives from the fact that the nucleus has a finite volume with surface effects playing a non-negligible role, in contrast to the effectively infinite volume of nuclear matter. Within the EOS, one can expect to be able to explore extreme regions where phase transitions like the one between a nuclear liquid phase and a nuclear gaseous

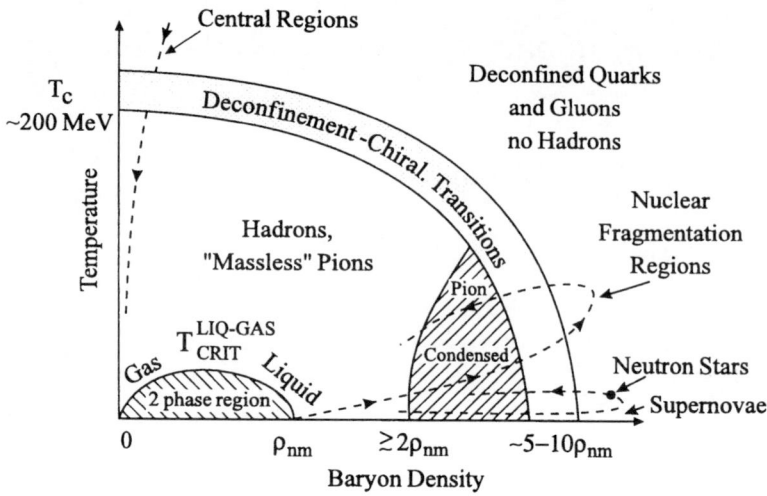

Fig. 5.1. Phase diagram of nuclear matter. The temperature is plotted versus the net baryon density for an extended volume of nuclear matter in thermal equilibrium. Normal nuclear matter appears at the point ρ_{nm} at zero temperature, and this is the neighborhood explored by low energy nuclear physics. The region of the phase diagram corresponding to quark deconfinement at a temperature T_C) and chiral symmetry restoration is indicated. Above T_C, hadrons dissolve into quarks and gluons. Above the temperature of chiral symmetry restoration, quarks are massless. The two critical temperatures may well coincide. The trajectories indicated in the phase diagram, show two paths for probing the quark–gluon plasma with high-energy nucleus–nucleus collisions. One involves reaching high baryon density among the hot, compressed fragments of the colliding nuclei, and the other involves very high temperatures where conditions may approximate those of the early universe. (Reprinted from *RHIC Report: Conceptual Design of the Relativistic Heavy Ion Collider* BNL-Report 52195. Brookhaven National Laboratory, with permission)

phase may occur. The study of these phase transitions provides a major challenge to heavy ion collisions at extremely high energies. When the heavy ions collide a large volume of the nuclear interior can be strongly heated with a corresponding increase in the energy density (a factor of 2–10). This occurs in a highly compressed state which is followed by an expansion phase. This collision process will give rise to particular flow patterns of nucleons that may yield important information about the EOS describing the inside of the nucleus. Of course, trying to connect these properties to the nuclear interacting many-body system, where one should start from a detailed knowledge of the nucleon–nucleon interaction as discussed in Chap. 2, is a very ambitious task. It will involve strong and decisive medium modifications that may change the nuclear liquid drop phase into totally different phases of matter. These will be discussed in more detail in Sect. 5.4

In the next three sections we discuss three imporant issues related to the creation of very dense and strongly heated forms of nuclear matter: (i) the fragmentation and multifragmentation of the projectile and target nuclei in the collision process; (ii) the various flow patterns that are formed in those collisions and that carry basic information on the nuclear EOS and about the coherence aspects in the nuclear medium; and (iii) the study of the phase transitions created by the heavy ion collision processes.

5.2 Nuclear Fragmentation and Multifragmentation

In heavy ion collisions, where energies of about 30–50 MeV/nucleon are reached, the conditions are such that the projectile velocity approaches the speed of the individual nucleons inside the target nucleus. The violent collisions will generate nucleons, clusters of particles, and heavier nuclear fragments and the evaporation spectrum can be used as a kind of 'thermometer' characterizing the nuclear temperature, T. Recent experiments have shown evidence that a nucleus can continue to absorb 'energy' up to about 6 MeV internal temperatures. Going above, the nuclear system can no longer contain the amount of energy and, when approaching the binding energy of individual nucleons, the atomic nucleus will start to break up into several individual fragments. It has been observed that, with increasing excitation energy, the number of particles evaporated also increases. It was originally thought that one could go on until the 'boiling' point of the nuclear liquid was reached, a situation where all nucleons will separate out of the bound nuclear regime (the process depicted in Fig. 3.1). The particular process of multifragmentation has shown clearly that, before reaching this boiling point, the nucleus changes its mode of deexcitation or cooling and starts breaking up into a number of fragments of which some can reach quite high mass values. The threshold for this multifragmentation process is well below the boiling point and thus it is probably impossible to reach the liquid–gas phase transition in a continuous way. This is not too unrealistic because nature needs less energy in order to cool a very hot nucleus into a number of fragments that are themselves rather tightly bound compared to a total break-up, i.e., a boiling off into all the individual protons and neutrons. So, the cooling and heating processes are balanced in such a way that optimal divisions and fragmentation occur for the particular internal energy content.

We illustrate this with the multiplicity distribution of mass fragments as a function of the incident energy for a central Kr on Au collision (Fig. 5.2). This multiplicity distribution goes through a maximum at E/A of about 100 MeV/nucleon. It is indicative of the fact that, below this point, large mass fragments are produced in great number, consistent with a mixture of a liquid with gaseous parts. At higher energies, the large abundance of smaller fragments points towards an increasing importance of a vapor system.

Describing these fragmentation and multifragmentation processes is very difficult since one needs to include both global and local properties in describing the distribution of quantum fluctuations that appear in a microscopic

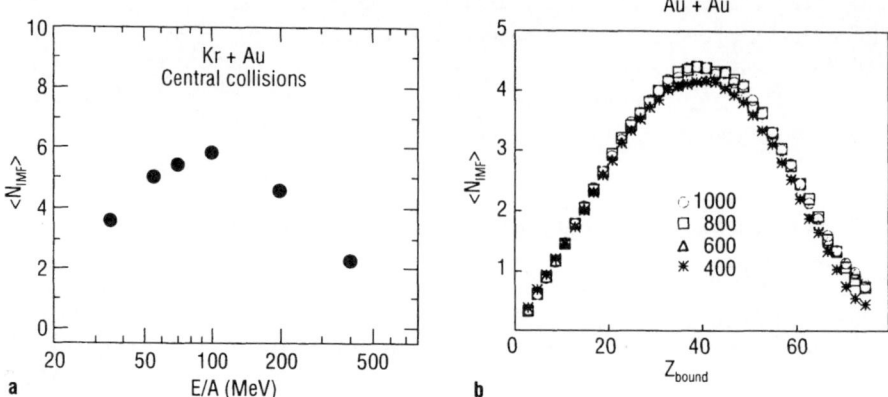

Fig. 5.2. (a) The solid points show the variation in the fragment multiplicities $\langle N_{\mathrm{IMF}} \rangle$ (to be more precise $\langle N_{\mathrm{IMF}} \rangle$ represents the mean value of the multiplicity (number) of fragments of intermediate mass per event), as a function of the incident energy in central Kr+Au collisions. The observed variation is consistent with a liquid–gas phase transition at around 100 MeV/nucleon. (Taken from NSAC (1996) *Nuclear Science: A Long Range Plan* February, with kind permission) (**b**) The intermediate mass fragment multiplicity distribution $\langle N_{\mathrm{IMF}} \rangle$ for Au+Au collisions with projectile energies of 400, 600, 800, and 1000 MeV/nucleon. The horizontal axis (Z_{bound}) denotes the sum of the nuclear charge numbers of all products from the decay of the projectile system, with the exception of hydrogen isotopes. (Taken from GSI-Nachrichten (1996) Vol. 5)

transport equation. Huge computer calculations based on nuclear relativistic hydrodynamics and Landau–Vlasov, Boltzmann–Uehling–Uhlenbeck (BUU) theories yield, at best, only very qualitative agreement with the experimental results in this energy domain of heavy ion collisions.

5.3 Flow Patterns
in Hot and Compressed Nuclear Matter

On reaching the much higher energy region of head-on collisions between heavy ions, approaching 1 GeV/nucleon, nuclear matter gets 'piled' up in the reaction zone. The participating nucleons cannot escape rapidly and a high-density strongly heated region of nuclear matter is formed. The way in which the momenta of incoming nucleons are transformed into transverse and longitudinal flow illustrates the high stopping power of the thus created interaction zone. Because the conditions that characterize this state of nuclear matter are typical for highly nonequilibrium systems, obtaining a full solution is a very difficult problem. It basically amounts to the calculation of the distribution of outgoing nucleons starting from the nucleon–nucleon force, taking into account its temperature behavior, and using the basic rules of quantum mechanics and relativity. This marriage of different elements results in a highly complicated problem. Its actual solution must be approximated by implementing specific models.

In one class of model, the atomic nucleus is handled as a liquid droplet and relativistic hydrodynamics is used to treat the collision process. The detailed behavior of individual nucleons is not followed but is averaged in a collective model. First large-scale studies of this type have been performed by Scheid, Müller, Hofmann, and Greiner (University of Frankfurt) [5.11] and also by Chaplin, Johnson, Teller, and Weiss at the Lawrence Livermore National Laboratory (LLNL) [5.12].

In another almost orthogonal approach, called the cascade model, nuclei in collision are described as a 'bag of marbles' (nucleons). The interaction of the two nuclei in the collision process is treated by calculating the trajectories of individual nucleons that are freely moving, except when collisions scatter a nucleon out of its original path, and where the collisions are described using the detailed knowledge of the nucleon–nucleon interaction in free space. This calculation thus simulates the paths in space of the individual nucleons. Cascade model calculations were first developed by Yariv, Fraenkel [5.14], and Cugnon [5.13].

Detailed comparisons show the basic differences between the two approaches. We discuss the results of both hydrodynamic and cascade model calculations for Nb–Nb collisions, as illustrated in Fig. 5.3. In the initial phase of the collision process, nuclear matter accumulates in a region not much larger than one of the separate nuclei. Explosive desintegration follows shortly after. The imporant differences between the two models only show up in the later stages of the heavy ion collision. In the cascade model, nuclei seem to be almost transparent for one another and most nucleons continue to travel quite close to the direction of the original heavy ion. In the hydrodynamic calculations, individual nucleons no longer appear but are put collectively in a fluid which, in the collision process, gives rise to a zone of greatly increased pressure, density, and temperature. A strong sideways emission of nucleons results, perpendicular to the original heavy ion trajectories. This latter sideways flow can be studied in more detail and is illustrated in Fig. 5.4 for the ^{40}Ar + ^{208}Pb collision at an energy of 800 MeV/nucleon. In such a collision, the projectile is moving at about 80% of the speed of light. The hydrodynamic calculations of Nix and Strottman [5.10] of Los Alamos National Laboratory and of Buchwald, Graebner, and Maruhn of the University of Frankfurt [5.4] show a developing shock way ahead of the nuclear matter becoming compressed in the subsequent collision. The resulting picture (Fig. 5.4) is quite spectacular in illustrating this very high energy process between atomic nuclei.

The verdict of nature on the question of whether heavy ions colliding are mainly stopped or appear almost transparent is given in experimental tests. Early experiments in a joint collaboration by Gutbrod, Stock at the

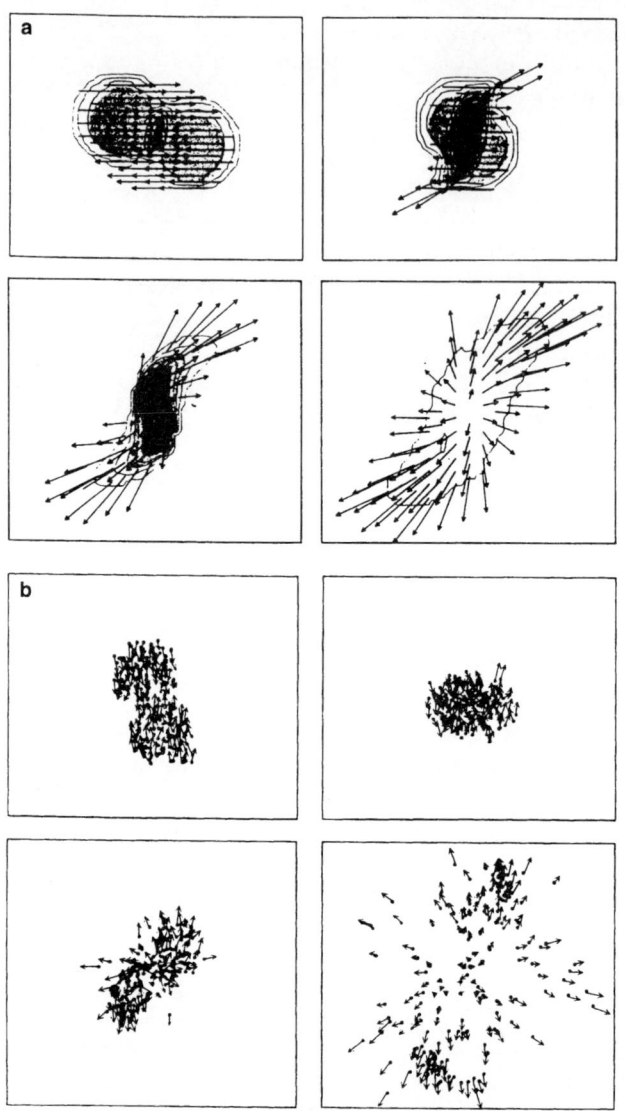

Fig. 5.3. In the upper part (**a**) a hydrodynamical model describing the heavy ·
ion collision between two Nb nuclei is illustrated. The collision is depicted in
the center-of-mass coordinate system so that the two nuclei seem to be moving
in opposite directions with equal speed. Significant matter flow perpendicular to
the original direction of motion is visible. Calculations have been carried out by
G. Buchwald, Ph. D. Thesis (1984) Univ. of Frankfurt, Unpublished and [5.4]).
The lower part (**b**) shows cascade model calculations for the same Nb+Nb heavy
ion collision process. Here, the path of each of the individual nucleon is calculated,
with only two-body interactions considered to govern motion. After an initial
state of compression, a nearly uniform angular distribution of fragments results.
Calculations have been carried out by J.J. Molitoris, Ph. D. Thesis (1985) Michigin
State Univ., Unpublished and [5.15]. (Taken from W. Greiner, H. Stöcker (1985)
Scientific American January, by courtesy of A.D. Iselin)

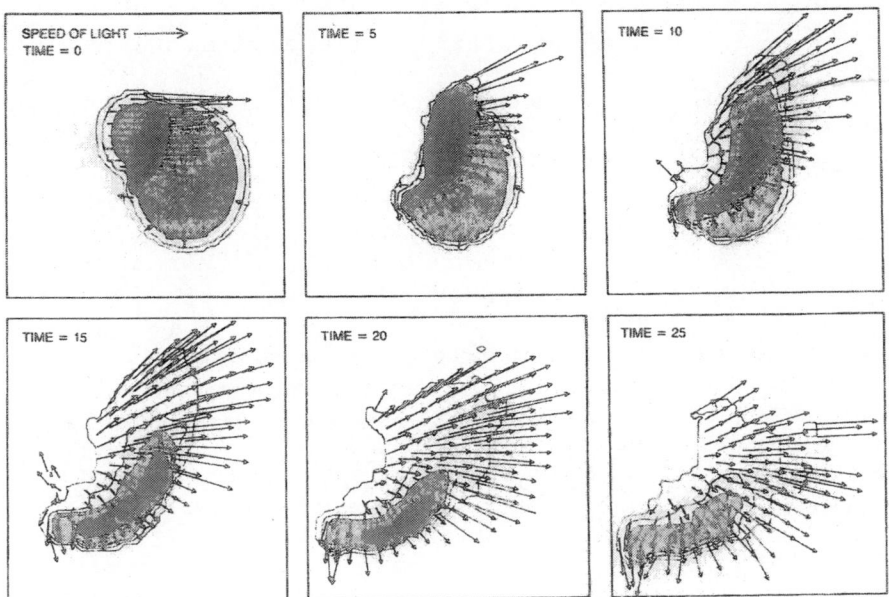

Fig. 5.4. Supersonic shock wave propagating in the collision of ^{40}Ar (coming from the left with an energy of 800 MeV/A) with a ^{208}Pb target. The collision is described using methods of relativistic hydrodynamics. The *arrows* represent the velocities of small volumes of nuclear fluid. The *contours* indicate isodensity lines. After an initial compression phase, a shock wave develops, first moving forwards and later sidewise. In the final stage, when the fused nuclei disintegrate, many fragments are emitted sideways. The unit of time corresponds to the time taken for light to travel one fermi (10^{-15}m). Calculations have been carried out by G. Graebner, Ph. D. Thesis (1984) Univ. of Frankfurt, unpubl. and [5.15]. (Taken from W. Greiner, H. Stöcker (1985) *Scientific American* January, by courtesy of A.D. Iselin)

GSI, Darmstadt and a team working at LBL headed by Poskanzer [5.4], [5.9] concentrating on high-multiplicity events that are the signature of head-on collisions, mainly seemed to confirm the sideways motion following the collision.

Unambiguous tests have been made more recently by Ritter, Gutbrod, and Poskanzer in a GSI–LBL collaboration [5.2]. These experiments used a magnificent apparatus called the PLASTIC ball, a spherical shell containing 800 plastic detector elements all surrounding the target position. This detector is unique in that it not only identifies the number of fragments but at the same time registers information about the mass, charge, and energy of each of the individual fragments. In collisions of ^{40}Ca on ^{40}Ca, the results were not conclusive but in similar collision experiments using ^{93}Nb on ^{93}Nb, results consistent with the outcome of hydrodynamical calculations were obtained. Even though cascade calculations come to the same conclusions for the light elements, the hydrodynamical model seems able to give the more generally valid predictions.

In some of the highest energy collisions carried out with the PLASTIC ball, with nuclei of energies as high as 2 GeV/nucleon, the data were consistent with a picture in which the initial projectile and target almost completely disappear, leaving a huge debris of particles and fragments that diverge from the center of mass, which is at rest. The analysis of the above collision process allowed the effective temperature in the collision zone to be deduced; it was as high as 1.5×10^{12} K (equivalent to 150 MeV). This is the highest temperature ever achieved under laboratory conditions. These experiments have enabled the EOS to be extrapolated to unknown regions.

So concluding on this part, it becomes clear that the collisions of heavy ions can supply us with important information about the behavior of compressed nuclear matter. This has direct bearing on such topics as the formation of neutron stars in the final stages of stellar evolution and on achieving a better understanding of the EOS of nuclear matter.

5.4 The Ultimate Phase Change: Quark–Gluon Phase Transitions

Upon further increasing the energy with which heavy ions collide and also increasing the volume (mass) of the colliding heavy ions so as to create a very hot, dense zone, one should not be surprised to find that the more 'classical' model descriptions (Sect. 5.1) fail to describe such violent collisions.

Classically, when heated, matter melts to become a liquid which can subsequently boil transforming the matter into a gaseous phase. These gas molecules can then, after further supply of heat (energy) become ionized forming a state of matter where the nuclei and electrons coexist in the form of a plasma.

It was speculated some time ago that nuclear matter, described as superfluid system, will at high enough temperature become vaporized (see Sect. 5.1 on fragmentation). A comparison of the corresponding phase diagrams for water and nuclear matter is carried out in Fig. 5.5. At a still later stage, it is thought that the "nucleon vapor" may even change into a plasma where the quarks and gluons coexist over a relatively extended region of space and time forming a 'quark–gluon' plasma. These early speculative ideas have in recent years been given a solid theoretical and experimental foundation. In this section we discuss the experimental and theoretical developments that led to the major results.

Before looking at some of the experimental efforts to reach the optimal conditions, we briefly consider why the quarks and gluons should actually form an 'unconfined' state at these very high energies and densities. One of the basic reasons can be found in the idea of charge screening which is also a

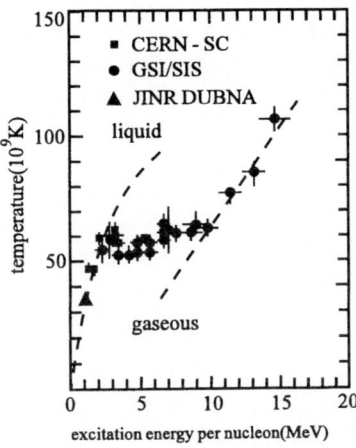
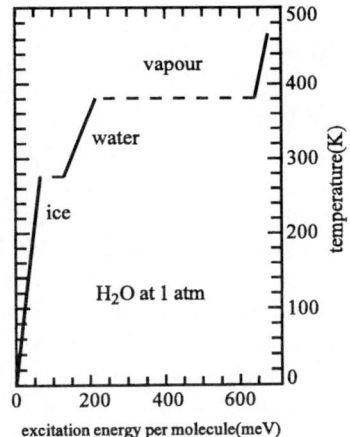

Fig. 5.5. Comparison of the phase diagrams [temperature T versus excitation energy per nucleon (in nuclear matter) or per molecule (for water)] for nuclear matter and water. The analogies are clearly evident. The "boiling" point at which, with increasing excitation energy per nucleon, the internal temperature remains essentially constant, is approximately $50-60 \times 10^9$ K . The data were taken at JINR (Dubna), CERN (SC) and, the more numerous results, at the GSI (Darmstadt). (Taken from GSI-Nachrichten (1996) Vol. 5)

well known idea in atomic physics. The force acting between charged systems in a bound-state configuration is influenced in an important way when many such systems are packed close together. The Coulomb force, e.g., is subject to screening in the presence of many other charges (Debye screening) and one obtains the expression for the screened potential

$$\frac{e^2}{r} e^{-r/r_{\mathrm{D}}} , \tag{5.1}$$

with r_{D} being the screening radius which is inversely proportional to the overall charge density of the system. If now the Debye screening radius becomes smaller than a typical atomic distance scale r_{Bohr}, the binding force between the atomic nucleus and the electrons becomes so well screened that the increased density may cause the system to change from an insulator to a conductor. This phase transition was first discussed by Mott [5.36]. Screening is thus a short-range mechanism that can dissolve the formation of bound states and one expects this to happen in the nuclear (bound) to quark–gluon (unbound) phase transition too. This new state would then be a conductor in relation to the basic charge, which is "color" charge in the theory of strong interactions or QCD.

In order to observe this phenomenon experimentally one must first understand as well as possible the energy deposition in head-on collisions of heavy ions with equal mass. Most experimental results come from hadron–hadron collisions, hadron–nucleus collisions, and the fixed target heavy ion collider experiments. The central feature in describing these collisions is the concept of 'nuclear transparency' which describes how the atomic nuclei interpenetrate in the high energy collision process. Combining all currently known data, the following picture is emerging (Fig. 5.6). At the low energy end the colliding nuclei bring each other to rest rather well. At the other, high energy, extreme the interaction process changes in an unexpected way and the two nuclei go through each other (they become transparent!) creating two hot, baryon-rich fireballs moving away from one another at high speed. We shall return to this process later. The energy at which transparency starts to set in, or the maximum energy for which the nuclei can still bring each other to rest, is estimated to be about 5–10 GeV/nucleon in the center of mass frame. It is precisely at this point that the energy density deposited in the fragmentation region becomes maximum. At these energies, when the nuclei pass through each other the nuclear volume has insufficient time to come to equilibrium, in contrast to the case of lower energy processes. For example, for two 100 GeV/nucleon Au nuclei colliding head on, the nuclei would literally pass through each other in such a way that roughly 90% of their energy remains in the nuclei with little change of direction. This leaves about 10% of the incident energy to excite the region of vacuum left around the center of mass. This central region, which is almost fully devoid of nuclear matter, can become excited to an energy density high enough that the "color" force of QCD gives rise to the conduction state described above. This in turn produces an abundance of mesons, baryon–antibaryon pairs. This much is fairly well established within the theory of QCD. In a more pictorial way this phenomenon is also known as 'melting' the vacuum.

Experiments carried out up to now have used beams of relatively light ions (^{16}O, ^{28}Si, ^{32}S) at both the AGS (Brookhaven National Laboratory) and at the SPS (CERN). In the experiments using a sulfur beam of 6.4 TeV or with about 200 GeV/nucleon at the SPS at CERN, an energy density of approximately 2 GeV/fm^3 has been reached. Such experiments are characterized by huge hadron multiplicities (see Fig. 5.7) and the formation of a 'freeze-out' zone (the region where the nuclei collided and where, after they have passed, the vacuum is heated so strongly that a quark–gluon plasma may form in which the entire "zoo" of particle physics may be recreated). The freeze-out zone in this case has twice the dimension of the projectile. Thus nuclear systems created in the laboratory can give rise to conditions that come close to the anticipated critical condition for the formation of a new phase of matter.

At this point we should also mention the signatures that would give an unambiguous indication of the formation of a quark–gluon plasma. This is not an easy subject because the transition between hot, compressed nuclear matter and the quark–gluon phase is still not well understood. A number of signals, however, are recognized as most probable indications for the formation of the quark–gluon phase of matter:

INITIAL STATE BEFORE COLLISION

(i)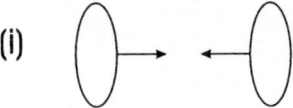

BARYONS STOPPED IN OVER-ALL CM

(ii)

AT HIGHER ENERGY, NUCLEI ARE TRANSPARENT TO EACH OTHER

(iii)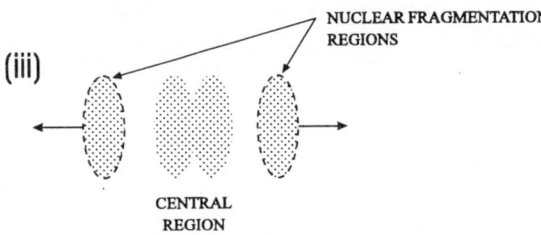

Fig. 5.6. Illustration of the various possible stages in a relativistic heavy ion (HI) collision: (i) Initial state before the collision. (ii) Depending on the available energy all baryons may be stopped in the overall center-of-mass system. (iii) At much higher energies, nuclei can become almost fully transparent to each other. In this process nuclear fragmentation regions are situated near the 'original' heavy ions but a highly heated (up to 10^{12} K) central region may be formed releasing quarks and gluons into a plasma configuration. (Reprinted from *RHIC Report: Conceptual Design of the Relativistic Heavy Ion Collider* BNL-Report 52195. Brookhaven National Laboratory, with permission)

1. The abundant formation of photons and leptons, which are not affected by the strong force, can carry information about the plasma state. Detailed studies of their spectra can track the thermal history and give information about the temperature and the duration of a phase transition.
2. The production of lepton pairs, technically called di-lepton production, contains information in their momentum spectra and in their yields. These observables are directly related to temperature. Unique signals will be difficult to find experimentally because such di-lepton pairs can be formed in quite a number of other ways which will form a large background.
3. A signal that will be most critical to detecting the setting in of a conducting state of quarks and gluons, and thus of "color" deconfinement, is the formation of heavy vector mesons, in particular the J/ψ particle. In a quark–gluon plasma, the quarks "lose contact" because they are in a conducting state and so have fewer interaction chances resulting in a decrease in the formation of, e.g., the J/ψ charmed quark–antiquark

Fig. 5.7. The many particle tracks produced at the CERN SPS accelerator, picturing the collision of a ^{32}S nucleus with a Au target at 200 GeV/A energies for the sulphur ions. Hundreds of charged fragments are produced, mostly protons and pions. (Taken from CERN Annual Report (1990) Vol. 1, with permission)

bound state. Of course, in order to obtain an unambiguous signal one has to incorporate all knowledge about J/ψ production in nucleon–nucleus collisions and in particular the absorption characteristics of J/ψ particles in nuclear matter.

4. If a plasma is formed for long enough that, besides thermal equilibrium, the equilibrium allowing the formation of large numbers of strange quark–antiquark pairs is also reached, in the cooling process, a large number of particles containing strange quarks could survive in the final state observed in the laboratory.

A number of other signals have been described and suggested but we shall not go into these technical details. In Table 5.1 we show, for completeness, a list of the various probes used in the search for a quark–gluon plasma.

There might of course be other unexpected results signaling the creation of this new state of matter. In the cooling process it would not be wild to suppose that the plasma may partially condense into strange forms of matter that do not belong to the normal world as we know it. As this matter will interact with the more standard particles appearing in our universe, unexpected phenomena may arise and be detected.

Just recently, preliminary results from the NA50 experiments studying J/ψ production in Pb–Pb collisions at an energy of 158 GeV/nucleon at CERN and comparing to former data obtained from the NA38/NA40 experiments seem to indicate that only about half of the originally created J/ψ

Table 5.1. Experimental probes of new states of matter. (Reprinted from *RHIC Report: Conceptual Design of the Relativistic Heavy Ion Collider* BNL-Report 52195. Brookhaven National Laboratory, with permission)

SIGNAL	COMMENTS	
Inclusive particle spectra Particle interferometry	Indicators of temperature size and density	*Global event parameters*
Multi-particle correlations in rapidity; Energy flow	Long range correlations and macroscopic fluctuations characteristic of first-order phase transition	
$J/\psi, \psi,' \Upsilon, \Upsilon'$ suppression in dilepton spectra	Color screening effects in deconfined plasma suppress heavy quark resonances	*Indicators of a phase transition*
Particle flavor ratios	Chemical equilibrium in hot plasma gives a large number of strange particles and enhanced $\bar{\Lambda}/\bar{p}$ ratio	
Stable multiquark states	Six-quark and higher configurations readily assembled in the plasma	
Direct photon production $(m_T = p_T)$	$m_T \leq 50\,\mathrm{MeV}$: coherent emission from local charge fluctuations $50 \leq m_T \leq 500\,\mathrm{MeV}$: hadronic decays; some coherent effects $500 \leq m_T \leq 3\,\mathrm{GeV}$: direct emission from plasma	*Penetrating probes: direct information from the plasma*
Lepton pair production (virtual photon: $m_T^2 = m_{\mathrm{Pair}}^2 + p_T^2$)	$m_T \geq 3\,\mathrm{GeV}$: approach to equilibrium; structure functions of quarks and gluons change and are computable in perturbative QCD	
High-p_T jets	Measures propagation of quarks and gluons through nuclear matter; hadronization properties reflect the "real sea" of quark–gluon plasma	

particles survive their journey through the interaction zone after the collision to the outside detectors (Fig. 5.8). Some theorists interpret this strong supression as an indication that many of the J/ψ particles are absorbed in a hotter-than-usual nuclear region which may also be indicative of the formation of a quark–gluon plasma phase, in at least one part of the colliding fireball. With the planned dedicated RHIC accelerator at BNL (see Box IX) which is on schedule and should give early results around 1998–1999 and, much further along the road, with the planning of Pb on Pb collisions at the planned LHC accelerator at CERN, opportunities to probe the vacuum and alter its properties will come within reach of physicists.

For example, it will most probably be possible to recreate conditions that might resemble the early phase in the creation of matter in our universe.

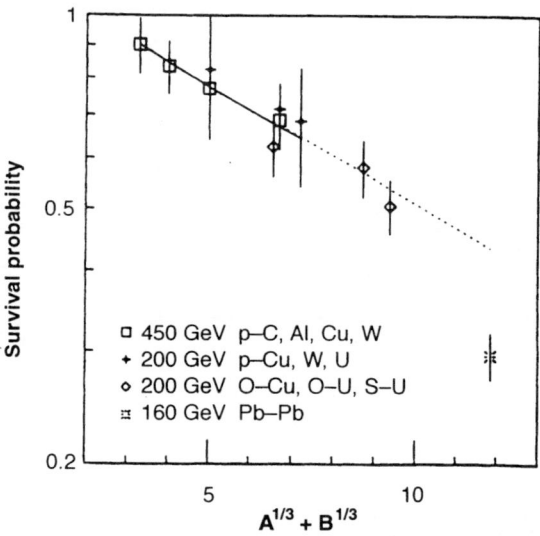

Fig. 5.8. The J/ψ survival probability, after absorption in nuclear matter, as a function of $A^{1/3} + B^{1/3}$, in which A and B correspond to the mass numbers of the colliding systems. The *full line* (*dashed-line*) is the survival probability for the proton–nucleus (nucleus–nucleus) systems. Experimental results are from the NA40/NA38 and NA50 collaborations. The abrupt change (in Pb–Pb collisions) could tentatively indicate the "melting" of a bound hadron into the quark–gluon plasma state. (Taken from J.P. Blaizot et al. (1996) Phys. Rev. Lett. **77** 1703. American Physical Society, with permission)

Box IX

RHIC: The Road to the Little Bang

In Brookhaven, in 1991, the construction of RHIC, a relativistic heavy ion collider began. This accelerator will occupy a 4 km circumference tunnel and will accelerate countercirculating beams of heavy ions, up to and including the very heavy Au nuclei, which fully stripped of all its electrons has a charge of 79^+. The energy associated with such an accelerated heavy ion will be as much as 100 GeV/nucleon!

This unique collider, which will become operational in 1999 if everything goes as planned, has as its major goal the creation and subsequent decay of a new state of matter: the quark–gluon plasma. Starting from basic theoretical concepts of QCD, a state of deconfinement of quarks and gluons could appear and lead to a color force conducting state of matter. In order to reach these extreme conditions, most probably those that governed the first milliseconds of our own universe, one should enable the largest aggregates of known nuclear matter (very heavy nuclei like Au and Pb) to collide head-on with the above energies. Thus far, such experiments have been carried out mainly at the AGS in Brookhaven National Laboratory and the SPS at CERN, where only fixed target experiments were possible until now. In those experiments glimpses of the strange behavior of nuclear matter when compressed and heated have been seen. Extrapolating onwards, some theorists have recently suggested the possible formation of clumps of the long awaited quark–gluon plasma starting from Pb+Pb collisions at CERN. It is clear though that a more dedicated apparatus is needed to allow an optimal study of these phenomena under controlled laboratory conditions. RHIC will consist of a double ring of superconducting bending magnets with only a modest bending field of 3.5 Tesla. The individual magnets each have a length of almost 10 m and one needs 1600 of them to complete the full set-up. There can be up to six interaction zones where the beams can be made to meet head-on creating the violent collision events. The beam tubes are much wider than in conventional proton or electron rings of comparable intensity (up to 8 cm to take into account repulsive Coulomb effects acting between the highly charged heavy ions). Moreover, the ions will be fully stripped when brought into the ring structure to attain the final energy. The machine will also allow asymmetrical collisions, e.g., between O and Au.

The starting point is formed by two existing tandem Van de Graaff accelerators. After passing these, the ion emerges partially stripped of its electrons at an energy of approximately 15 times its net charge. A 700-m long transfer line will then transport the accelerated ions into the newly built booster which can accelerate ions as heavy as Au and produce energies of almost 100 MeV/nucleon. Subsequently, these ions will pass through a stripper channel that will remove all but the innermost electrons: these too will be stripped off for the heaviest elements. The ions then enter the AGS accelerator which increases the energy towards 28 GeV for protons and 10 GeV/nucleon for

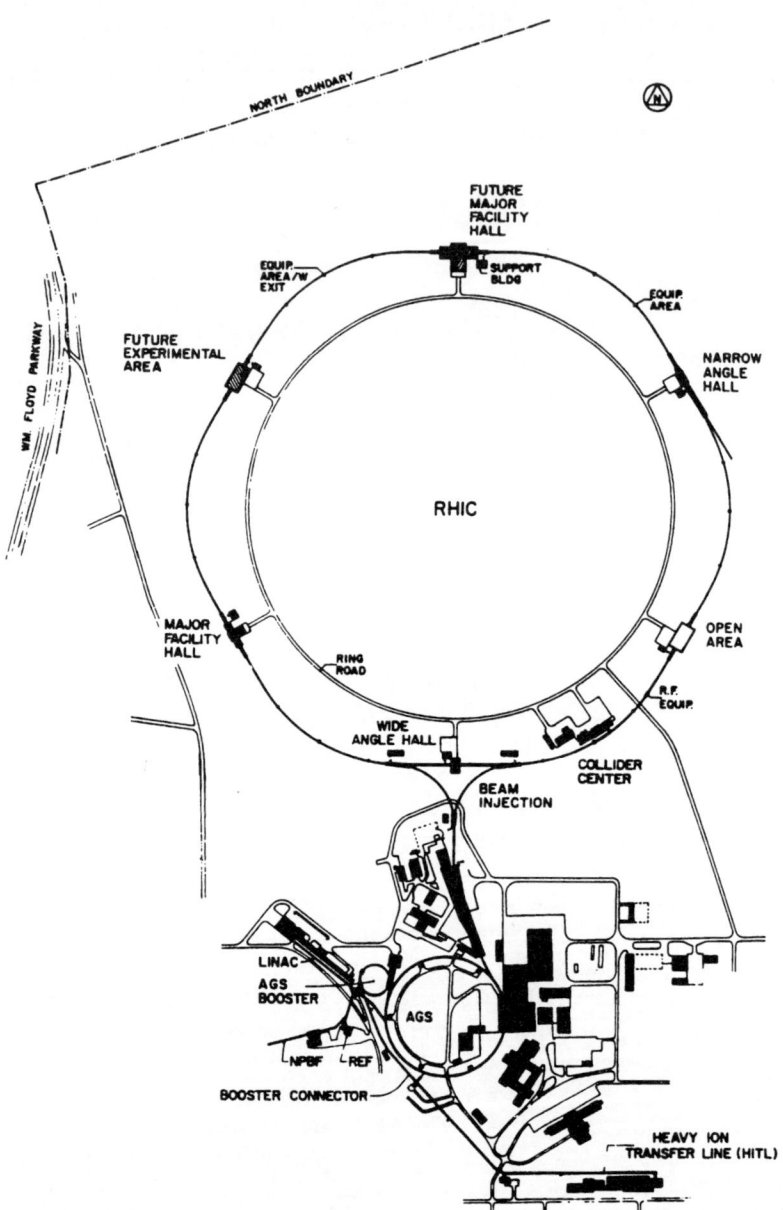

Fig. IX.1. The ultrarelativistic heavy-ion collider (RHIC) under construction at the Brookhaven National Laboratory. The rounded, hexagonal curves are the heavy-ion storage rings carrying two countercirculating beams of heavy ions around the 4 km circumference. They can be made to collide at six interaction points. The source–injector systems will use the existing Van de Graaff and AGS accelerators before injecting ions into the RHIC rings. (Reprinted from *RHIC Report: Conceptual Design of the Relativistic Heavy Ion Collider* BNL-Report 52195. Brookhaven National Laboratory, with permission)

Au ions. Then one last stripper will remove any remaning electrons and prepare the ions to be brought into the RHIC accelerator. It then only takes a short time to produce and shorten bunches of ions and accelerate them up to the final energy, while keeping them countercirculating across the various interaction (crossing) points on the ring. The entire layout is illustrated in Fig. IX.1. A number of advanced collider detectors have been conceived: two major systems called PHENIX and STAR and two smaller ones, BRAHMS and PHOBOS. They are under construction at present. The detailed characteristics will not be described here, but suffice to say that they provide complementary capabilities and will immediately go into use when RHIC is first started up. A number of initial experiments are already planned and installed with the following aims:

1. To obtain first results on the energy densities, the multiplicity densities, temperatures, etc., reached, so as to determine the degree of equilibration in the Au + Au collisions.
2. To obtain first results on the spacetime evolution of matter under these extreme energy densities and to study the presence of a possible long-lived mixed phase.
3. To study properties related to chiral symmetry.
4. To gather data on the suppression of the J/ψ particle as a function of the transverse energy for the Au+Au collisions where one expects initial energy densities about one order of magnitude larger than those achieved in experiments with a fixed target.
5. To measure the particles emitted with large transverse momentum.

In addition to these heavy ion collider experiments, detailed programs of proton–proton and proton–nucleus scattering will be performed. These complementary experiments will provide a good understanding of background processes that might otherwise obscure analysis of the quark–gluon plasma signals. These many experiments will involve interdisciplinary teams of nuclear and particle physicists.

In conclusion, RHIC will provide nuclear physics with a new and unique apparatus to study the ultimate structure of nuclear matter under extreme conditions of temperature, pressure, and density. Its most important contribution may well turn out to be something totally unexpected by way of information about new forms and unexplored regions of matter.

5.5 Further Reading

Studies of the equation of state (EOS) using heavy-ion collisions are reported in a large body of references. Here we refer to a book devoting extensive discussion to the equation of state.

5.1 Greiner, W., Stöcker, H. (eds.) (1989) *The Nuclear Equation of State, Part A: Discovery of Nuclear Shock Waves and the EOS* (Plenum, New York)

An instructive popular article introducing this topic is:

5.2 Gutbrod, H., Stöcker, H. (1991) Scientific American, November, p. 32

Technical papers that give a good overview are:

5.3 Ainsworth, T.L., Baron, E., Brown, G.E., Cooperstein, J., Prakash, M. (1987) Nucl. Phys. **A464**, 740
5.4 Stöcker, H., Greiner, W. (1986) Phys. Rep. **137**, 277

The study of fragmentation and multifragmentation with a view to under-standing the liquid-to-gaseous phase transition in the nucleus has been pur-sued intensively and much experimental effort has been made to identify this phase transition. We mention two popular accounts and a technical report.

5.5 Ngô, C. (1988) Clefs CEA 11, Hiver (in French)
5.6 GSI Nachrichten (1996) ed. by Braun-Munzinger, P., Emling, H., Gross, K.-D., Kluge, H.-J., Lantzch, J. Vol. 5, 13 (in English)
5.7 Ogilvie, C.A. et al. (1991) Phys. Rev. Lett. **67**, 1214

Heavy ion collisions have been performed with the major aim of discover-ing the interaction processes and how incoming energy and momentum are dissipated, giving rise to the flow patterns characteristic of these HI collisions.

First we give two popular accounts:

5.8 McHarris, W., Rasmussen, J.O. (1984) Scientific American, January, p. 44
5.9 Greiner, W., Stöcker, H. (1985) Scientific American, January, p. 58

A review paper by Nix, a number of the original papers, some more recent reports on Au+Au collisions are:

5.10 Nix, J.R. (1979) Progr. Part. Nucl. Phys. **2**, 237

5.11 Scheid, W., Müller, H., Greiner, W. (1974) Phys. Rev. Lett. **32**, 741
5.12 Chaplin, G.F., Johnson, M.H., Teller, E., Weiss, M.S. (1973) Phys. Rev. **D8**, 4302
5.13 Cugnon, J. (1980) Phys. Rev. **C22**, 1885
5.14 Yariv, Y., Fraenkel, Z. (1979) Phys. Rev. **C20** 2227 and (1981) Phys. Rev. **C22**, 488
5.15 Molitoris, J.J., Stöcker, H. (1985) Phys. Rev. **C32**, 346

5.16 Baym, G., Monien, H., Pethick, C.J., Ravenhall, D.G. (1990) Phys. Rev. Lett. **64**, 1867
5.17 Bondorf, J.P., Botvina, A.S., Mishustin, I.N., Souza, S.R. (1994) Phys. Rev. Lett. **73**, 628

5.18 Gaimard, J.J., Schmidt, K.-H. (1991) Nucl. Phys. **A531**, 709
5.19 Hsi, W.C. et al. (1994) Phys. Rev. Lett. **73**, 3367
5.20 Schnedermann, E., Heinz, U. (1992) Phys. Rev. Lett. **69**, 2908

One of the major issues in developing still higher heavy ion accelerators is finding unambiguous signals for the creation of a new state of matter: a quark–gluon plasma phase. This topic, together with many details of HI processes and the behavior of matter under extreme conditions of pressure, temperature and density, has also been discussed in a large number of popular accounts:

First we give a couple of books containing many further references to the literature:

5.21 Creutz, M. (1983) *Quarks, Gluons and Lattices*, Cambridge Monographs on Mathematical Physics (Cambridge University Press, Cambridge)
5.22 Greiner, W., Stöcker, H. (eds.) (1989) *The Nuclear Equation of State, Part B: QCD and the Formation of the Quark– Gluon Plasma* (Plenum, New York)

There is a fairly extensive set of conference proceedings going by the name Quark Matter. We refer to some of the earlier ones.

5.23 Quark-Matter-89 (1990) Nucl. Phys. **A498**
5.24 Quark-Matter-90 (1991) Nucl. Phys. **A525**
5.25 Quark-Matter-91 (1992) Nucl. Phys. **A544**
5.26 Quark-Matter-93 (1994) Nucl. Phys. **A566**
5.27 Quark-Matter-95 (1995) Nucl. Phys. **A590**
5.28 Quark-Matter-96 (1996) Nucl. Phys. **A610**

There are also many popular accounts, including

5.29 Baym, G. (1985) Physics Today, March, p. 40
5.30 Crawford, H.J., Greiner, C.H. (1994) Scientific American, January, p. 58
5.31 Goldhaber, J. (1992) LBL Research Review 17, No. 2, 10
5.32 Physics News Update (1996) No. 289
5.33 Satz, H. (1986) Nature, Vol. 324, November, p. 116

Some interesting review articles and a number of more specialized articles concentrating on important issues in detecting a quark–gluon plasma are:

5.34 Gross, D.J., Pisarski, R.D., Yaffe, L.G. (1981) Rev. Mod. Phys. **53**, 43
5.35 Kajantie, K., Mc.Lerran, L. (1987) Ann. Rev. Nucl. Part. Sci. **37**, 293
5.36 Mott, N.F. (1968) Rev. Mod. Phys. **40**, 677
5.37 Satz, H. (1985) Ann. Rev. Nucl. Sci. **35**, 245
5.38 Shuryak, E. (1980) Phys. Rep. **61**, 72

5.39 Bertsch, G., Brown, G.E. (1989) Phys. Rev. **C40**, 1830
5.40 Blaizot, J.P., Ollitrault, J.-Y. (1996) Phys. Rev. Lett. **77**, 1703
5.41 Brown, G.E., Bethe, H.A., Pizzochero, P.M. (1991) Phys. Lett. **B263**, 337
5.42 Gavin, S., Satz, H., Thews, R.L., Vogt, R. (1994) Z.Phys. **C61**, 351
5.43 Heiselberg, H., Pethick, C.J., Staubo, E.F. (1993) Phys. Rev. Lett. **70**, 1355
5.44 Jacob, M. (1995) Nucl. Phys. **A583**, 13
5.45 NA 50 coll., Bordalo, P. et al. (1996) in: *Proc. of Rencontres de Moriond*, ed. by Tranh Tan Van, J. (Edition Frontières, Gif-sur-Yvette)
5.46 Seibert, D. (1989) Phys. Rev. Lett. **63**, 136

To close this chapter, we give some references pertaining to the RHIC, its construction, and its physics:

5.47 Brown, G.E., Oset, E., Vicente Vacas, M., Weise, W. (1989) Nucl. Phys. **A505**, 823
5.48 RHIC Report (1989) *Conceptual Design of the Relativistic Heavy Ion Collider* BNL-Report 52195 (Brookhaven National Laboratory, Upton, N.Y.)
5.49 Schwarzschild, B. (1991) Physics Today, August, p. 17

The latest update information about the progress of the RHIC can be obtained from the RHIC Bulletin, BNL, Upton, New York 11973-5000.

6. The Nucleus as a Laboratory
for Studying Fundamental Processes

6.1 Introduction

Because of its structure of protons and neutrons which, in the bound nucleus, are sensitive to the strong, weak, and electromagnetic interactions, the atomic nucleus can serve as a very specific "laboratory" for testing how these basic interactions behave together. It should be possible to probe physics at the intersections of nuclear physics with fields as varied as particle physics (tests of particle properties, tests of the standard model, ...), atomic physics, quantum physics and astrophysics (see Fig. 6.1). The very important issue of nuclear astrophysics will be discussed in much more detail in Chap. 7.

It will become clear that under certain conditions the atomic nucleus may act as a 'magnifier' of important aspects that would otherwise be difficult to study. The study of neutrino properties is an example: Neutrinos can be studied in great detail near the end-point of beta decay and the consequences of neutrino properties can be deduced from experimental work on double beta decay.

Before going in more detail and refraining from an exhaustive list, one should mention aspects that

(i) pertain in large part to tests of the standard model and to searches for new physics beyond it, and,
(ii) study the consequences of even tiny violations of the conservation laws in physics.

In the next sections we shall concentrate on those nuclear physics tests that carry important information about neutrino properties (beta decay, double beta decay, intrinsic neutrino properties), formulate some general conclusions about the role of nuclear physics in this very important domain of tests of the 'fundamentals' of physics, and, finally, discuss symmetry tests for the invariances of the basic interactions (T violation, P violation, electric dipole moment of the neutron, etc.).

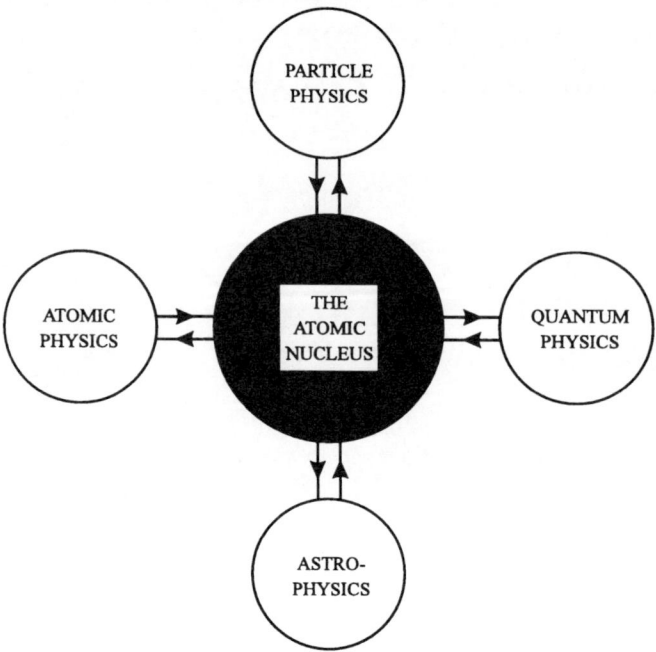

Fig. 6.1. Schematic illustration showing the important position occupied by the atomic nucleus in relation to fundamental experiments in atomic physics, astrophysics, and particle physics, as well as for testing basic concepts in quantum physics.

6.2 Beta Decay and Double Beta Decay: A Road to the Neutrino Mass

In studies of the weak interaction, which causes a proton to be transformed into a neutron, a positron, and a neutrino, or a neutron into a proton, an electron, and an antineutrino, detailed experiments, measuring near the end-point of the electron spectrum, e.g., in tritium beta decay, have given upper limits for the neutrino mass. The probability of finding an electron with a given energy E_e in beta decay exhibits a very specific dependence on the neutrino mass $m_{\bar{\nu}_e}$, on the total energy released in the beta decay process, E and on the electron energy, E_e in the spectrum of electrons emitted in this beta decay process, given by the form

$$\propto \sqrt{1 - \frac{m_{\bar{\nu}_e^2} c^4}{(E - E_e)^2}} \, . \tag{6.1}$$

This has allowed detailed tests of energy spectra near the point where the electron carries away almost all the energy available in the decay process (Fig. 6.2). Because the energy release (called the Q-value in technical terms)

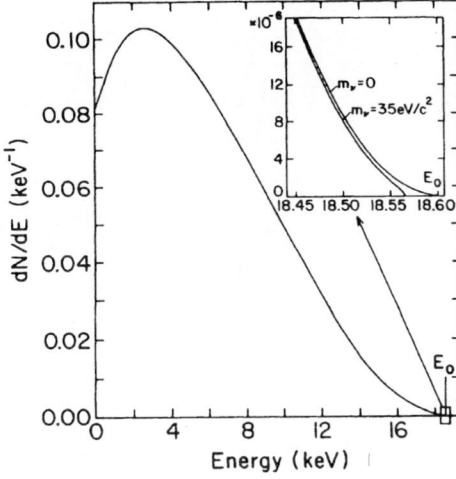

Fig. 6.2. The tritium (^{3}H) beta spectrum, assuming an end-point energy of $E_{e(\text{max})} = 18.6\,\text{keV}$. The *insert* illustrates two possible scenarios corresponding to vanishing neutrino mass and a small but non-vanishing (35 eV/c^2) neutrino mass. (Taken from Holzschuh (1992) Rep. Progr. Phys. **55**, 1035. IOP Publishing, with permission)

in the beta decay of tritium is only 18.6 keV, this nuclear beta decay process has been studied in great detail. There has been quite some controversy about these measurements and subsequent experiments by Bergkvist and Lubimov were in disagreement about the neutrino mass (Fig. 6.3). The more recent data all seem to be consistent with an upper limit for the electron antineutrino mass of 10 eV.

Situations can also arise where the mass of a nucleus is such that the regular beta decay cannot occur but a second-order process, called double beta decay, is possible. This process which proceeds through the intermediate, virtual, states for the second-order process (see Fig. 6.4 for a schematic view) can be described in second-order perturbation theory by the expression

$$\langle f|H_{\text{int}}|i\rangle = \sum_n \frac{\langle |fH_{\text{int}}|n\rangle\langle n|H_{\text{int}}|i\rangle}{E_i - E_n} \;, \tag{6.2}$$

and a very crude estimate for the timescale of these processes is given by the square of a typical single beta decay matrix element.

In its simplest form double beta decay can be depicted as the second order process (Fig. 6.5a)

$$^A_Z X_N \rightarrow ^A_{Z+2} Y_{N-2} + e^- + e^- + \bar{\nu}_{e^-} + \bar{\nu}_{e^-} \;, \tag{6.3}$$

with the emission of two electrons and two electron antineutrinos. A few parent nuclei that might decay via this mechanism are ^{48}Ca, ^{96}Zr, ^{100}Mo, ^{116}Cd, ^{124}Sn, ^{130}Te, ^{150}Nd, and ^{238}U.

The lifetime estimates are typically of the order of $T_{1/2} \simeq 10^{20}$ y for an energy release of 2.5 MeV and this double beta decay has been observed by geochemical techniques through the extraction of the $Z + 2$ daughter nuclei out of an ore of the original Z mother nucleus (Table 6.1) .

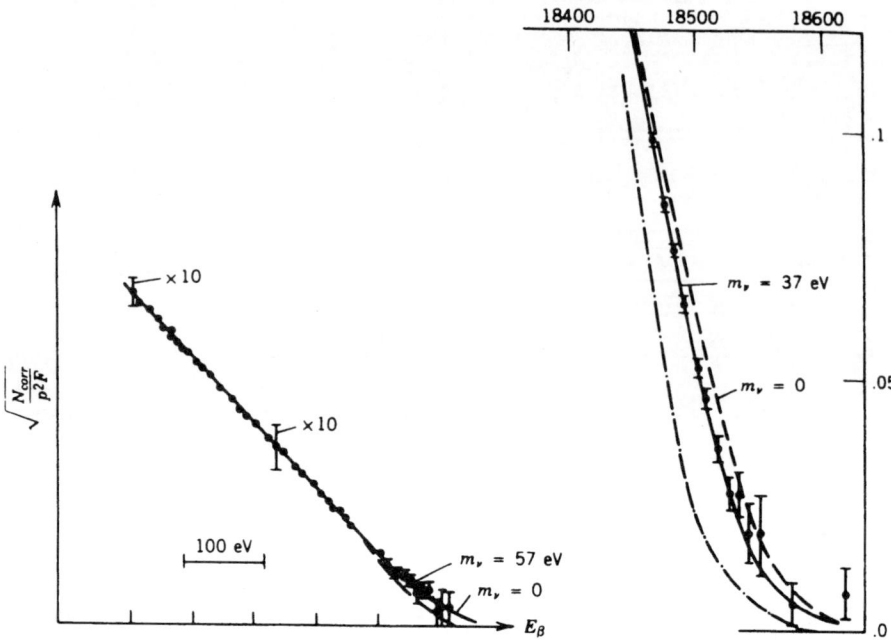

Fig. 6.3. Fermi-Curie plots for the tritium (^{3}H) beta decay. The data on the left-hand side are from Bergkvist and are fully consistent with a vanishing neutrino mass, and exhibit an upper limit of 57 eV. The more recent data from Lubimov (*right-hand side*) seem to favor a slightly non-zero neutrino mass of $\simeq$ 30 eV. (Adapted from Bergkvist et al. (1972) Nucl. Phys. **39**, 317 (*left part*), and Lubimov et al. (1980) Phys. Lett. **94B**, 266 (*right part*). Elsevier Science, NL, with kind permission)

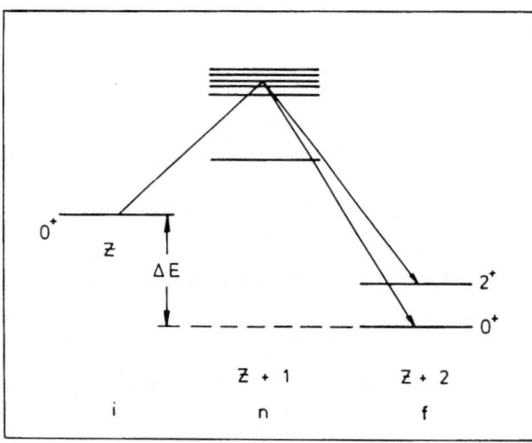

Fig. 6.4. A simplified set of level energies for three consecutive nuclei (Z, $Z+1$ and $Z+2$) where single beta decay from $Z \rightarrow Z+1$ is impossible. In this situation, a double beta decay process, proceeding through a set of intermediate (virtual) levels in the nucleus $Z+1$, is possible. (Taken from K. Heyde *Basic Concepts in Nuclear Physic* ©1994 IOP Publishing, with permission)

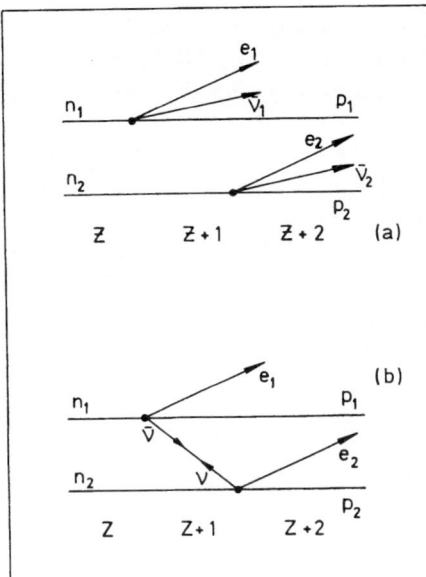

Fig. 6.5. Schematic illustration of the double beta decay process in which two neutrons are transformed into two protons, two electrons, and two electron antineutrinos (**a**). In the lower part (**b**), neutrino-less beta decay is shown. Here, in simple language, one of the emitted electron antineutrinos is absorbed as an electron neutrino, transforming the neutron into a proton and an electron. (Taken from K. Heyde *Basic Concepts in Nuclear Physic* ©1994 IOP Publishing, with permission)

Table 6.1. Summary of selected double beta decay results. (Taken from K. Heyde *Basic Concepts in Nuclear Physic* ©1994 IOP Publishing, with permission)

	Experiment Geochemistry	Laboratory	*Calculation* Doi et al. [6.8]	Haxton et al. [6.10]
^{76}Ge				
$T_{1/2}$ (2ν)(y)			2.3×10^{21}	3.7×10^{20}
$T^{1/2}$ (0ν)(y)		$> 3.7 \times 10^{22}$	9.4×10^{22}	
m_ν (eV)			< 16	< 7
^{82}Se				
$T_{1/2}$ (2ν)(y)	1.5×20^{20}	$(1.0 \pm 0.4) \times 10^{19}$	1.5×10^{20}	1.7×10^{19}
$T^{1/2}$ (0ν)(y)		$> 3.1 \times 10^{21}$	3.2×10^{22}	
m_ν (eV)			< 33	< 12
^{130}Te				
$T_{1/2}$ (2ν)(y)	2.6×10^{21}		2.6×10^{21}	1.7×10^{19}
$T^{1/2}$ (0ν)(y)			2.5×10^{22}	
m_ν (eV)			< 130	
$^{130/128}$Te				
$T_{1/2}^{130/128}$	$(1.0 \pm 1.1) \times 10^{-4}$			
m_ν(eV)			< 5	< 5

There are, however, more exotic ways to describe double beta decay processes. Within certain forms of Grand Unified Theory (GUT) there is no strict conservation law for the leptons and so neutrino-less beta decay could be possible (Fig. 6.5b). Laboratory experiments to discover the actual way in which double beta decay processes occur are of fundamental importance in nuclear physics and far beyond. Figure 6.6 shows the energy spectrum in double beta decay in the regular case and for the non-standard neutrino-less beta decay. All has to do with whether the neutrino is a Majorana or a Dirac particle (charge conjugate state to the original neutrino is identical (different) to (from) the original neutrino itself) or, expressed in symbols, whether

$$C|\nu_{e^-}\rangle \equiv |\bar{\nu}_{e^-}\rangle = |\nu_{e^-}\rangle(M) \ ,$$

or

$$C|\nu_{e^-}\rangle \equiv |\bar{\nu}_{e^-}\rangle \neq |\nu_{e^-}\rangle(D) \ . \tag{6.4}$$

The beta decay particle balance then appears as follows

$$^A_Z X_N \rightarrow ^A_{Z+2} Y_{N-2} + e^- + e^- \ , \tag{6.5}$$

with the creation of just two electrons. This distinction between a Majorana or Dirac nature was tested a long time ago in the Davis experiment where a clear cut difference between the electron neutrino and electron antineutrino was proven. Davis thus concluded that the electron neutrino was not a Majorana particle. The demise of parity conservation in the weak interaction by the ground-breaking experiment of Wu changed this early interpretation. It was shown that neutrinos and antineutrinos were particles with different helicity (different screw type) so the right-handed antineutrino could never match a left-handed neutrino

$$n \rightarrow p + e^- + \bar{\nu}_{e^-}(R) \ ,$$
$$\nu_{e^-}(L) + n \rightarrow p + e^- \ . \tag{6.6}$$

Thus, the double beta decay process would be strictly forbidden even if the electron neutrino was a Majorana particle. These statements are only strictly true, however, if the electron neutrino and antineutrino are massless particles. Therefore, the double beta decay experiment, carried out under strict laboratory conditions, and the determination of the character of the neutrino, become of very basic importance.

The first full laboratory experiment studying the double beta decay process was performed by Elliott, Hahn, and Moe and for the nucleus ^{82}Se which decays into ^{82}Kr giving clear-cut evidence for the emission of two electrons during the decay process. More recent experiments using enriched ^{76}Ge detectors have come up with limits close to 1 eV for electron neutrinos that have Majorana character. The precise analyses depend on the theoretical calculation of a number of nuclear properties but ongoing experiments should enable an improvement of the upper limit by a factor of 5. A summary of existing experimental and theoretical results for double beta decay processes is given in Table 6.1.

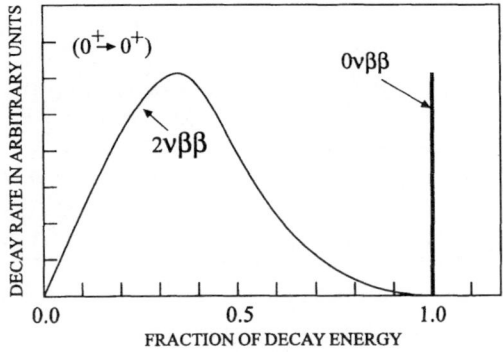

Fig. 6.6. The two-electron energy spectrum for a $0^+ \rightarrow 0^+$ double beta decay process. Distinction is made between the two-neutrino and neutrino-less situations, denoted by $2\nu\beta\beta$ and $0\nu\beta\beta$, respectively. (Taken from K. Heyde *Basic Concepts in Nuclear Physic* ©1994 IOP Publishing, with permission)

6.3 The Elusive Neutrino

Almost 40 years after their first discovery, the basic properties of the neutrinos (at present three families have been observed: the electron-, muon-, and tau-type neutrinos), remain largely unknown. We do not yet know if the neutrino is distinct from its antiparticle, nor whether it has a non-zero rest mass. A large number of experiments have provided strict upper limits (see Sect. 6.2) but no lower limit has yet been set.

Besides the major production of neutrinos in nuclear reactions (through the beta decay of the fission products, which are neutron rich and transform into more stable elements producing electron antineutrinos), neutrinos are amply produced in the solar burning process where protons fuse into heavier elements (deuterium, helium, ...), as will be discussed in Chap. 7 in more detail. Solar neutrinos provide a unique way to study the internal solar physics. However, the neutrino–nucleus cross-sections are only of the order of 10^{-43} cm^2 making it very difficult to register these neutrinos in detectors on earth, despite the very large neutrino flux at the earth's surface (about 10^{11} cm^{-2} s^{-1}).

Two important facts concerning neutrino processes and their detection are:

(i) solar neutrinos have been observed under controlled laboratory conditions;
(ii) the number of detected neutrinos is substantially lower than predicted from solar models.

In order to discuss what these two experimental facts imply about neutrino physics we need to consider:

(i) the neutrino production mechanisms in the fusion reactions in the sun;
(ii) the detection in the various types of detectors available;
(iii) possible implications for the intrinsic neutrino properties;
(iv) planned future experiments.

6.3.1 Neutrino Production Mechanisms

The various reactions that produce neutrinos in the solar interior are summarized in Table 6.2. The energy spectra for these different reactions showing the relative contributions to the observed neutrino flux are illustrated in Fig. 6.7.

Table 6.2. Solar neutrino production reactions. (Taken from K. Langanke, C.A. Barnes ©1996 Adv. in Nucl. Phys., Vol. 22. Plenum Publishing Corp., with permission)

Reaction	Maximum E_ν (MeV)	Flux (10^{10} cm^{-2}s^{-1})
$p + p \rightarrow {}^2H + e^+ + \nu$	0.42	6.1
$p + e^- + p \rightarrow {}^2H + \gamma$	1.44	1.5×10^{-2}
${}^7Be + e^- \rightarrow {}^7Li + \nu$	0.86 (50%)	4.0×10^{-1}
	0.38 (10%)	
${}^8B \rightarrow {}^8Be + e^+ + \nu$	14.06	4.0×10^{-4}
${}^{13}N \rightarrow {}^{13}C + e^+ + \nu$	1.20	5.0×10^{-2}
${}^{15}O \rightarrow {}^{15}N + e^+ + \nu$	1.73	4.0×10^{-2}

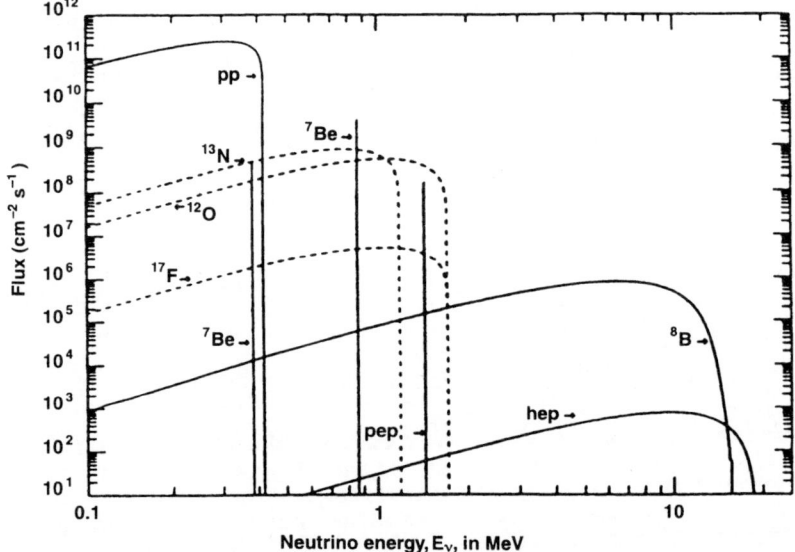

Fig. 6.7. The energy spectrum of solar neutrinos, as predicted from the standard solar model. The *solid lines* correspond to neutrinos produced in pp-chains, *dashed lines* to neutrinos produced in the CNO-cycle (see Chap. 7). The various processes are discussed in more detail in the text and also in Table 6.2. (Taken from Oberauer et al. (1992) Rep. Progr. Phys. **55**, 1093. IOP Publishing, with permission)

It is seen that the most basic reaction in which two protons are fused into a deuterium nucleus, producing a positron and an associated electron neutrino, gives by far the largest contribution but the neutrinos produced in this reaction have an upper energy limit of 0.42 MeV. At the other extreme, neutrinos produced in the beta decay of ^{8}B are the only ones that have energies exceeding 2 MeV and go up to about 16 MeV but their production rate is very low compared to the p+p fusion reaction.

6.3.2 Neutrino Detection and the Missing Neutrino Problem

All of the detectors used to register neutrinos on earth have certain thresholds: if one wants to detect as many neutrinos as possible, in particular those producing the largest contribution to the total flux, one has to use reactions that have as low a threshold as possible, preferably below 0.42 MeV. The neutrino capture rates in the various experiments are so tiny that they are measured in the unit SNU (for Solar Neutrino Unit) which corresponds to 10^{-36} captures per second and per target atom. The main detectors are:

(i) In the Homestake mine (South Dakota), which was the first dedicated neutrino detector and started data taking in 1970, one uses the inverse electron capture reaction on ^{37}Ar, i.e.,

$$\nu_e +^{37} Cl \rightarrow e^- +^{37} Ar \,, \tag{6.7}$$

with a threshold at 0.814 MeV. So this reaction is sensitive mainly to the higher energy neutrinos produced in the beta decay of ^{8}B. Only about 32% of the predicted neutrino flux is detected in this experiment.

(ii) The Kamiokande detector, originally conceived to detect the possible decay of the proton, has been taking most interesting neutrino data from the 1987a supernova and from the sun. This detector is a Cerenkov water detector and thus uses the scattering of electron neutrinos off electrons. The threshold is rather high at about 5 MeV. This detector is exclusively sensitive to the ^{8}B decay neutrinos, which it has been detecting since 1985.

(iii) The Gallex detector is a huge detector made of 30 tons of gallium. It detects neutrinos through the reaction

$$\nu_e +^{71} Ga \rightarrow e^- +^{71} Ge \,, \tag{6.8}$$

with a threshold at 0.233 MeV. This detector was constructed primarily to respond to the major source of neutrinos, i.e., the p+p fusion reaction. It is situated in the Gran Sasso tunnel in Italy and has been taking data since 1991. Gallex detects about 60% of the predicted neutrino flux.

(iv) The Sage experiment uses the same reaction as in Gallex but with 60 tons of metallic gallium. The detector is located in a mine in the Caucasus mountains and has been taking data since 1989. In this experiment the neutrino detection rate is, at most, 56% of the predicted value.

One general conclusion is that, compared to the standard solar model predictions of Bahcall, there is a striking deficiency in the neutrino detection rate. The problem so created is a serious one because the reaction cross-sections producing the neutrino flux inside the sun are well known. Measurement of the production rates of those elements that are at the origin of neutrino emission is of major importance. One of those reactions, the $^7\text{Be}(\text{p},\gamma)^8\text{B}$ reaction (radiative proton capture), is very critical and efforts are under way to measure the production cross-sections corresponding to solar temperatures (energies) using radioactive ion beams. More details will be discussed in Chap. 7.

6.3.3 Neutrino Mass and Neutrino Oscillations

The problem of the missing neutrinos could be solved by assuming that a number of the electron neutrinos that induce the reactions measured in the various detector set-ups change their character into other flavors of neutrinos. This assumption is not as wild as it might seem because, a non-zero but even tiny mass implies 'neutrino-oscillations' between the three different flavors and so could explain the missing neutrino problem and ascribe a non-zero mass to the electron neutrino at the same time.

A number of explanations for the missing solar neutrinos have been put forward; all imply physics beyond the standard model of particle physics, and all imply that some of the neutrino species are not massless! We shall not discuss the details of the suggested solutions simply referring to the vast literature that exists on neutrino physics. We concentrate here on the idea that the physical neutrino states are not identical to the neutrino eigenstates that correspond to the free Hamiltonian.

Within a two-state model, considering only the electron and muon neutrino types (ν_e and ν_μ as the physical states, and ν_1, ν_2 the corresponding mass eigenstates), as in ordinary quantum mechanics, a state prepared in a given configuration at a given initial time t_0 can evolve according to the time-dependent Schrödinger equation. If we denote the physical neutrino eigenstates at time $t = 0$ as linear combinations of the mass eigenstates:

$$|\nu_e\rangle = \cos\theta_v|\nu_1\rangle + \sin\theta_v|\nu_2\rangle$$
$$|\nu_\mu\rangle = -\sin\theta_v|\nu_1\rangle + \cos\theta_v|\nu_2\rangle \,, \tag{6.9}$$

with a mixing angle θ_v, the time evolution results in states at time t, given by

$$|\nu_e\rangle_t = \cos\theta_v e^{-iE_1 t}|\nu_1\rangle + \sin\theta_v e^{-iE_2 t}|\nu_2\rangle$$
$$|\nu_\mu\rangle_t = -\sin\theta_v e^{-iE_1 t}|\nu_1\rangle + \cos\theta_v e^{-iE_2 t}|\nu_2\rangle \,, \tag{6.10}$$

where E_1 and E_2 are the masses of the two mass eigenstates $|\nu_1\rangle$ and $|\nu_2\rangle$. It is now easy to evaluate the probability that an electron neutrino, which starts evolving at time $t = 0$, is transformed into a muon neutrino at time t.

It is given by

$$|\langle \nu_\mu | \nu_e \rangle|^2 = \sin^2 2\theta_v \sin^2 (\pi x / L_v) \ , \tag{6.11}$$

where x is the distance traveled by the neutrino and L_v the "vacuum neutrino oscillation" length which itself is given by the expression

$$L_v = \frac{4\pi}{m_1^2 - m_2^2} \ . \tag{6.12}$$

In this way the neutrino flavor can be modified. Several experiments have been carried out to search for such oscillations without unambiguous success although some hints for oscillations between muon antineutrino and electron antineutrino flavors starting from positive muon decay at rest have recently been found by the Los Alamos LSND collaboration.

The current status seems to indicate that neutrino oscillations in vacuum between the site of production (the interior of the sun) and the detector on earth would need a rather implausible fine tuning of the various quantities appearing in (6.11), i.e., of the quadratic mass difference, the mixing angle, the travel of distance x, and the average neutrino energy. This severe constraint does not seem to be necessary if neutrino oscillations are assumed to take place mainly in the solar interior (including interactions with matter).

Mikheyev, Smirnov, and Wolfenstein (MSW) [6.31], [6.32] have developed a model which can explain in a rather natural way the combined Ga, Cl, and Kamiokande neutrino detection results. In this approach, which is discussed at length by Langanke and Barnes [6.24], Oberauer and von Feilitsch [6.25], and Van Klinken [6.45], an electron neutrino can change into a muon neutrino by a resonant interaction.

Even for small mixing angles for oscillations in vacuum, an effective conversion of the neutrino spectrum can occur because the electron density inside the sun changes with radial distance from the center. Near the solar centre, the electron neutrino is primarily in the state $|\nu_2\rangle$ but at the solar surface, it again reaches its vacuum value. A global solution from the MSW model implies a small Δm^2 value of the order of $\leq 10^{-4}$ eV^2 indicating that electron and muon neutrinos should have almost equal mass. This is illustrated in Fig. 6.8 and is evidence that the MSW mechanism can simultaneously explain the observed neutrino fluxes for the various detectors currently in operation. In this figure, the upper left part is excluded because of accelerator, reactor, and atmospheric observations. Contours for fixed SNU values show MSW interpretations compatible with the various detector experiments. The full black regions are allowed by all three detectors.

The various detectors now operational all suffer from very low event rates and poor resolution. Therefore, a number of very ambitious direct counting detectors are planned, which should provide a final solution to the longstanding problem of the elusive neutrino.

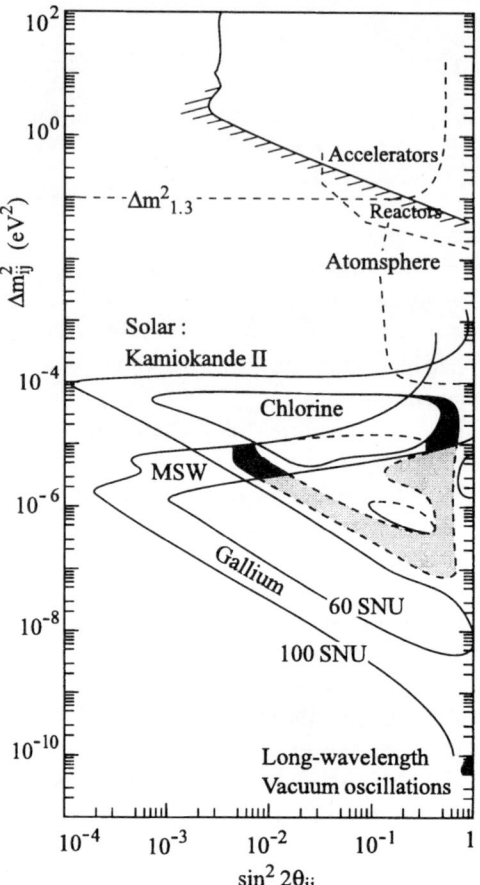

Fig. 6.8. Bound regions on neutrino oscillations in Δm_{ij}^2 and $\sin^2 \theta_{ij}$. The *top right* region is excluded by accelerator, reactor, and atmospheric observations. Iso-SNU lines for solar neutrinos indicate the MSW interpretation of solar-neutrino observations: The area between *full lines* is allowed by the Ga experiments, between the *dashed lines* by the Cl experiments, and that between the *dotted lines* by Cerenkov experiments. The *black zones* denote regions allowed by all three detectors (at a 90% confidence limit). At the *bottom*, there remains a small region for a long-wavelength interpretation of solar neutrinos. (Taken from J. van Klinken (1996) J. Phys. G **22**, 1239, IOP Publishing, with permission)

6.3.4 Planned New Neutrino Detector Experiments

A number of new detectors are currently being constructed.

(i) The Sudbury Neutrino Observatory (SNO) in Canada is a detector using heavy water where deuteron dissociation induced by neutrinos is the reaction detected. The two possible reactions in this case are

$$\nu_e + d \rightarrow p + p + e^-$$
$$\nu_L + d \rightarrow p + n + \nu_L \,, \tag{6.13}$$

where the first reaction is induced by the charge changing interaction and can only be induced by electron neutrinos, whereas the second, the neutral interaction, can be initiated by all three types of neutrino. It now all depends on the spectrum of neutrinos appearing at the detector: If the MSW mechanism is indeed active, then the various species of neutrino should be present in the incident spectrum and the neutral reaction should dominate over the electron neutrino induced type. The final state $(p+p+e^-$ versus $p+n+\nu_L)$ should make it easy to identify the reaction that actually takes place and so an unambiguous signal should appear from this experiment. Needless to say, everybody is anxiously awaiting the start of this experiment.

(ii) Super-Kamiokande, an upgraded version of the earlier Kamiokande experiment, sensitive to the high energy ^8B neutrinos, should have sensitivity that is sufficiently improved to observe possible distortions of the electron spectrum at low energy.

(iii) The Borexino detector, under construction in the Gran Sasso tunnel, will be constructed from liquid scintillator material and will be sensitive to the flux of low-energy neutrinos produced in electron capture reactions on ^7Be.

(iv) A number of deep underwater detection set-ups are planned using megatons of H_2O, see [6.45].

With all these experiments starting up soon, detailed data on the precise neutrino spectral distributions will become available and will enable stringent tests of the solar model. The SNO experiment, which has been planned as the ultimate trap forcing the neutrino to reveal its true mass nature, will allow us, in the coming years, to finally get to grips with the neutrino. It will at the same time yield answers to deep questions relating to the problem of missing mass within the universe. A possible solution to the 'dark matter' problem may well be found at SNO within a couple of years!

6.4 The Essentials of the Neutron: Famous for 14 Minutes 49 Seconds

The free neutron decays into a proton, an electron, and an electron antineutrino with a half life of 14 m 49 s, an observable that has been determined to rather good precision only very recently. Accurate measurements of various properties of the neutron: half-life, decay asymmetries, electric dipole moment, magnetic properties, etc., have an important bearing on the precision of the standard model of particle physics.

A precise determination of the half-life of the free neutron has been a long-term effort and only quite recently, in 1990, Byrne, Dawber and co-workers succeeded in obtaining a really precise value of 893.6 ± 5.3 s (the variation in measurement of this half-life over the last 35 years is illustrated in Fig. 6.9). The team working at the ILL, Grenoble detected not the electron final decay product but the protons. Because the protons, as decay products, are relatively slow moving and have charge $+e$, they can be trapped in a device called a Penning trap where the combination of electric and magnetic fields holds them in this 'electromagnetic bottle' to be counted. In the neutron

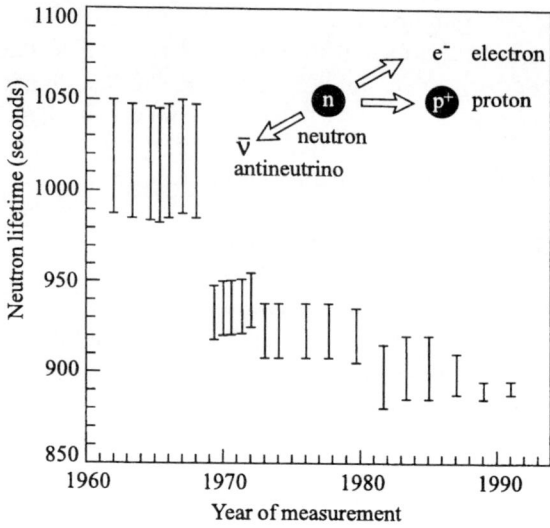

Fig. 6.9. The variation in the measured lifetime ($T_{1/2}$) for a free neutron over a period of more than three decades. Only very recently could rather precise values be obtained (see also text). (Adapted from J. Gribbin (1993) *New Scientist* March 13, ©IPL Magazines, with permission)

beam at the ILL, about 4×10^9 neutrons passed through the 'bottle' every second leaving behind just 10 protons from the radioactive decay.

Another method based on the decay of ultra-cold neutrons measures the number of surviving neutrons rather than the number that decay. The observed lifetime was in quite good agreement with the Penning trap measurement. In 1992 the accepted half-life was 889.1±2.1 s.

The neutron half-life, the various angular and polarization correlation coefficients in free-neutron beta decay, as well as log ft values (log ft value is a reduced beta decay half-life and separates out the energy dependence into a comparative half-life) in Fermi superallowed $0^+ \rightarrow 0^+$ beta decay all constitute some of the most stringent and best tests for the V-A (vector/axial-vector current) structure of the weak beta decay process. In the case of the now very precise neutron half-life, limits on possible departures from the standard model can be set. The neutrino half-life is also related to the primordial deuterium and helium content in the universe. This has to do with the fact that in a Universe at temperatures of around 10^{10} K, protons are continuously changing into neutrons (beta decay and inverse electron capture by protons were in statistical equilibrium). With the cooling of the universe, this process departs from equilibrium and neutrons start to decay one-sidedly into protons, electrons, and electron antineutrinos. At temperatures near to 10^9 K the neutrons start sticking to protons to form primordial deuterium nuclei with the strong force overcoming thermal dissociation. Thus, at about a time of 200 s after the formation of the universe, most neutrons got stuck in deuterium nuclei and thence in He nuclei. These processes and the subsequent primordial deuteron and ^{4}He content are highly sensitive to the neutron half-life. Calculating in the reverse direction, from the observed values of those primordial abundances, a neutron half-life of close to 890 s was derived, much lower than that previously measured, even in the second half of the 1980s.

Another most interesting result from the precise neutron half-life combined with the value of the primordial ^{4}He-to-proton ratio is a prediction of the number of neutrino families. The outcome was a value of 2.6±0.3 for the number of neutrino families, a prediction that has been advancing the experimental result on the number of neutrino families as obtained recently at LEP from the decay width of the Z^0 particle.

The neutron, first identified by James Chadwick at Cambridge in 1932, has many important things to say about a wide spectrum of physics. We refer to a number of more detailed papers and a recent book concentrating at great length on the many properties and characteristics of the neutron [6.36]–[6.39].

6.5 Breaking of Fundamental Symmetries: Low-Energy Tests

Symmetries have always been a basic guiding principle along the road to a better understanding of fundamental interactions and in searching for unification schemes in physics. Detailed tests of the various symmetries we expect to find in describing the laws of physics are therefore of the utmost importance. Discovering symmetry-breaking effects is actually an indication that we do not yet fully understand the most elementary processes and the way to describe them. The experimental fact, that on the level of the weak force (e.g. beta decay), left–right symmetry no longer holds was what enabled identification of the rules unifying the electromagnetic and weak forces into the electroweak theory. We shall discuss here some tests of time-reversal breaking as well as of the detection of nuclear parity violating effects.

6.5.1 Tests of Time-Reversal Invariance

In general, one expects that on a microscopic scale the particles that are influenced by the basic forces and governed by a set of equations of motion, will not be affected by a change in the way that time runs (backward or forward direction).

The only well-known case where time-reversal symmetry is broken (usually called T violation) has been obtained from the study of the decay of the neutral kaon particle. A deeper understanding of what causes time symmetry to be broken bears on our understanding of important questions such as the baryon asymmetry as it appears in the particle content of our universe. Specific nuclear and particle tests for T violation can be made in the study of the possible existence of electric dipole moments in the neutron, electron, and certain nuclei. Detailed searches for the electric dipole moment in the neutron over the years, with ever-increasing sensitivity since 1950, have been giving more and more stringent tests of the theoretical description of such T-violating effects in the basic theories. We illustrate the increasing sensitivity limits in Fig. 6.10.

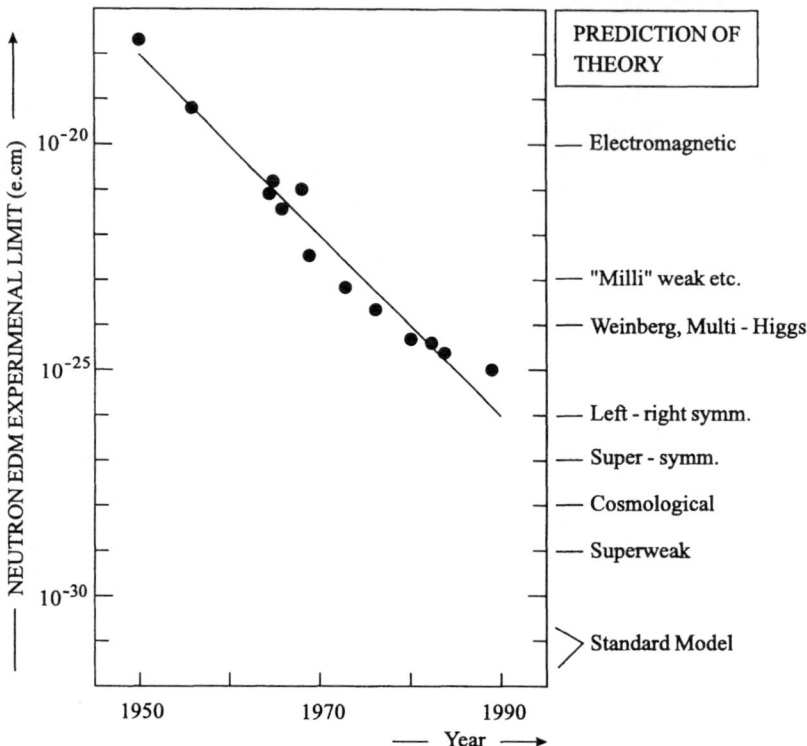

Fig. 6.10. A compilation of the results of the continuing search for a non-vanishing electric dipole moment for the free neutron. The various predictions stemming from theoretical models and constraints are indicated on the *right-hand side* of the figure. (Taken from NUPECC Report (1991) *Nuclear Physics in Europe: Opportunities and Perspectives* November, with permission)

   These experiments require ultra-high precision rather than of the use of large accelerators. Here, the increasing ingenuity of experimental research at small accelerator facilities in university laboratories in isolating and manipulating even a single atom is apparent. Using electromagnetic trapping methods, and cooling atoms with lasers, one is able to determine the basic properties of atoms and particles with enormous precision.

6.5.2 Tests of Parity Violation in Nuclear Physics

The major goal here is to determine the strength of the coupling constant for that part of the weak nucleon–nucleon force component which breaks parity as a good quantum number. Within the nuclear many-body system, a number of effects can play an amplifying role enabling one to study the relative role of the strong and weak force components inside the atomic nucleus.

Starting from the full Hamiltonian containing a large parity-conserving part and a very small parity non-conserving (PNC) contribution

$$H = H_0 + V_{\mathrm{PNC}} ,$$
(6.14)

lowest order perturbation theory leads to a mixing in the parity of nuclear eigenstates given by the relation

$$\psi_{J+} = \phi_{J+} + \sum_{J-} \frac{\langle \phi_{J-}|V_{\mathrm{PNC}}|\phi_{J+}\rangle}{E_+ - E_-}|\phi_{J-}\rangle$$

$$= \phi_{J+} + \epsilon\phi_{J-} .$$
(6.15)

Here, the typical strength of the PNC matrix elements is of the order of 1 eV and energy denominators are of the order of 1 MeV and so mixing amplitudes are of the order of 10^{-6}.

A favored testing ground for parity violating effects are very close-lying parity doublets in light nuclei. Here, one needs in addition very precise information about nuclear wavefunctions in order to single out the precise PNC effects.

There are a number of types of experiment that are particularly interesting for testing PNC:

(i) The study of cases where an observable would vanish exactly were it not for PNC. The alpha decay of unnatural parity states to 0^+ groundstates like $^{16}\mathrm{O}(2^-) \rightarrow {}^{12}\mathrm{C}(0^+) + \alpha$ gives direct access to the parity-violating matrix element;

(ii) Interference between PNC and PC matrix elements in the determination of pseudoscalar observables. A common variable is the circular polarization $P_\gamma = \langle \boldsymbol{\sigma}_\gamma.\boldsymbol{p}_\gamma \rangle$ of gamma rays emitted by unpolarized nuclei. As we discuss in a separate technical box (Box X), an amplification factor arises that greatly facilitates the determination of the circular polarization in light nuclei.

(iii) Interference between PNC and PC matrix elements for scalar observables. As a result, the observable is of second order in the weak component and so can be neglected in the first instance when studying PNC effects in a nuclear medium.

The detailed and precise alpha decay branching in $^{16}\mathrm{O}(2^-)$ leading to $^{12}\mathrm{C}(0^+)$ directly gives the PNC decay width of $(1.03 \pm 0.28) \times 10^{-10}$ eV.

The study of interference effects (ii) is particularly interesting because here one detects gamma radiation with very good statistics. In Box X we

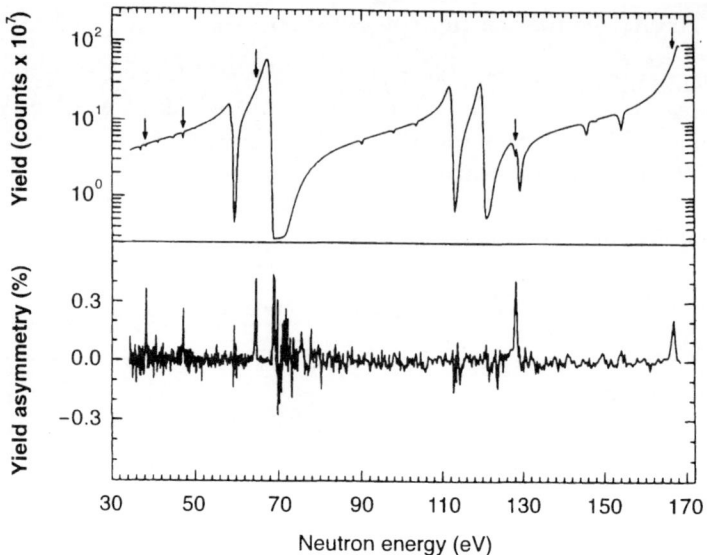

Fig. 6.11. The *upper part* of the figure illustrates the transmission of neutrons through the target of ^{232}Th. The large dips correspond to well-known resonances. The *lower part* plots the difference in transmission between left- and right-handed polarized neutrons. The strikingly large differences at certain energies correspond to weak resonances which, however, violate parity (up to $\simeq 10\%$). The experiments were carried out at Los Alamos National Laboratory. (Taken from NSAC (1996) *Nuclear Science: A Long Range Plan* February, with kind permission)

shall discuss in some detail the parity mixing effect occurring in light nuclei as deduced from the measurement of circular polarization.

Other interesting parity-violating effects can be studied by performing scattering experiments, which can be of the following types:

(i) scattering of polarized electrons or,
(ii) scattering of polarized protons off the proton, and,
(iii) polarized neutron scattering off atomic nuclei.

The quantity that is measured in all of these cases, the asymmetry A_{exp} is defined by the expression

$$A_{\mathrm{exp}} = \frac{\sigma^+ - \sigma^-}{\sigma^+ + \sigma^-} \quad , \tag{6.16}$$

where $+$ and $-$ denote the helicity of the polarized projectile that is scattered off the unpolarized target.

In the scattering of polarized cold neutrons at Los Alamos, extraordinarily large parity-violating effects have been observed. Even though the fractional parity violation in the nuclear strong force is only about one part per million, effects on the order of $1/10$ have been observed and so large enhancement effects in the nuclear medium must be at work. The results for the transmission of neutrons on ^{232}Th are shown in Fig. 6.11 (upper part) where the large dips are caused by a number of well-known resonances.

The lower part shows the scattering asymmetry and the large differences at certain energies are caused by weak resonances which lead to parity-violating effects of the order of 10%. The mechanisms causing these very big enhancements are not yet well understood.

Similarly, the p-wave scattering of polarized neutrons on ^{139}La, in which the number of transmitted neutrons indicates a strong p-wave resonance at 0.75 eV, shows a particularly large scattering asymmetry indicating parity-violating effects (Fig. 6.12).

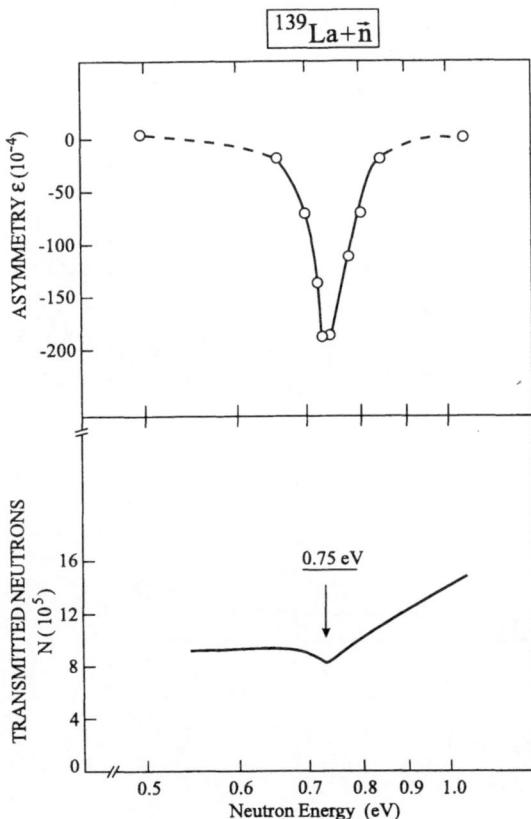

Fig. 6.12. Number of transmitted neutrons showing a p-wave resonance in the reaction ^{139}La $+n$ (*lower part*) and the corresponding asymmetry (*upper part*), indicating a sizable P-violating effect. (Reprinted from A. Richter (1993) Nucl. Phys. **A553**, 417c. Elsevier Science, NL, with kind permission)

These large effects can be understood in terms of the fact that, near threshold, neutron widths (penetrabilities) are proportional to the factor $(kr)^{2l+1}$. Through the parity-violating component in the strong force, coupling between p and s resonances can result, and this gives rise to the PNC observable being proportional to $[\Gamma_n(s)/\Gamma_n(p)]^{1/2} \propto 1/kr$. In the particular case of ^{139}La $+$

n, the amplification factor for detecting the PNC amplitude itself is almost 760. Experiments by Alfimenkov et al. [6.47] were able to derive a PNC matrix element of the order of 1×10^{-3} eV. A detailed but schematic model explaining the amplification factor in this case has been discussed by Richter (see caption of 6.12).

Box X

Parity Violation in Light Nuclei

In light nuclei, one can find cases where two levels are very close-lying, allowing parity mixing to be detected through the observation of circular polarization of gamma radiation deexciting these states. A number of good cases – ^{14}N, ^{18}F, ^{19}F, and ^{21}Ne – are shown in Fig. X.1.

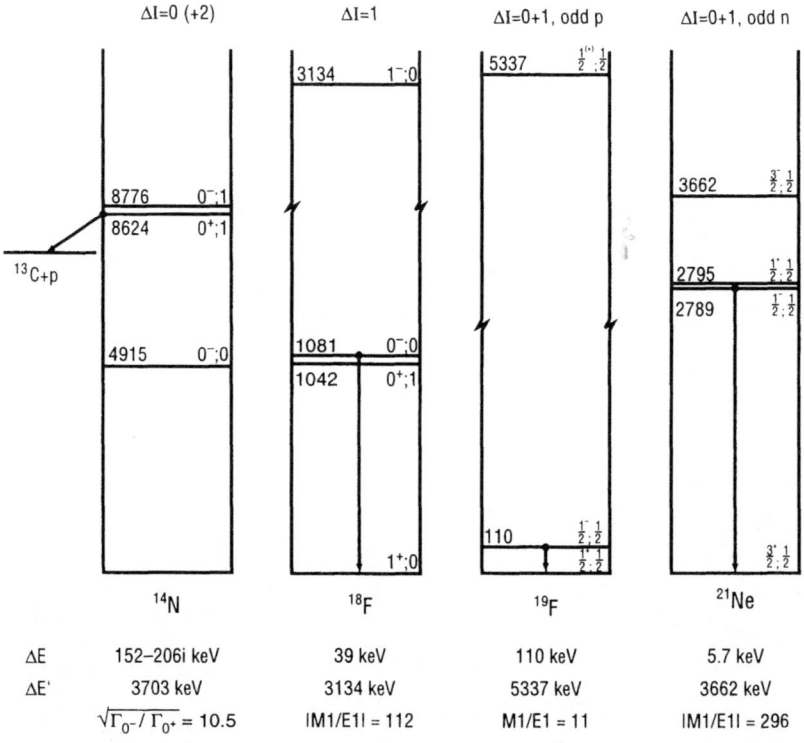

	^{14}N	^{18}F	^{19}F	^{21}Ne
ΔE	152–206i keV	39 keV	110 keV	5.7 keV
$\Delta E'$	3703 keV	3134 keV	5337 keV	3662 keV
	$\sqrt{\Gamma_{0^-}/\Gamma_{0^+}}$ = 10.5	IM1/E1I = 112	M1/E1 = 11	IM1/E1I = 296

Fig. X.1. Parity-mixed doublets in various light nuclei. The particular transitions displaying the amplified parity non-conservation are shown in each case. The numbers, denoted by ΔE, are the smallest energy denominators governing the parity mixing amplitude using perturbation theory. The numbers below indicate the various "amplification" factors. (Reprinted from the *Annual Report of Nuclear & Particle Science* ©1985, Vol. 35. Annual Reviews Inc., with permission)

In the particular case of ^{18}F, one has a 0^+ and a 0^- level with an energy separation of $\Delta E = 39$ keV and an excitation energy of about 1 MeV. From first order perturbation theory and defining the PNC mixing amplitude ϵ as

$$\epsilon = \frac{\langle+|V_{\mathrm{PNC}}|-\rangle}{39\mathrm{keV}} \, , \tag{X.1}$$

one obtains for the parity mixed wave functions

$$|1081\rangle = |-\rangle + \epsilon|+\rangle$$
$$|1042\rangle = |+\rangle - \epsilon|-\rangle \, , \tag{X.2}$$

and the resulting value for the circular polarzation variable caused by a small admixture of the 0^+ state into the 0^- level at 1081 keV becomes

$$
\begin{aligned}
P_\gamma &\cong 2\,\mathrm{Re}\left\{ \frac{\epsilon\langle gs|M1|+\rangle}{\langle gs|E1|-\rangle} \right\} \\
&\cong \frac{2}{39}\,\mathrm{Re}\left\{ \langle+|V_{\mathrm{PNC}}|-\rangle \frac{\langle gs|M1|+\rangle}{\langle gs|E1|-\rangle} \right\} \\
&\cong \frac{2}{39}\,\mathrm{Re}\left\{ \langle+|V_{\mathrm{PNC}}|-\rangle \left(\frac{\tau_-}{\tau_+}\right)^{1/2} \times \left(\frac{1081}{1042}\right)^{3/2} \right\} \, .
\end{aligned}
\tag{X.3}
$$

The amplification factor then becomes nearly 112 (the sign, however, cannot be determined by this method). From the data on ^{18}F, the absolute value of the PNC matrix element emerges as $|\langle V_{\mathrm{PNC}}\rangle| \leq 0.09$ eV. Theory (Haxton, [6.42]) gives a result of 0.37 eV.

6.6 Further Reading

We begin by referring to the references of Chap. 1 which include a number of books concentrating on general nuclear physics. There, beta decay is presented in some detail. Because beta decay is a particularly important issue a number of books completely devoted to this topic, for example, are given:

6.1 Holstein, B. (1989) *Weak Interactions* (Princeton University Press, Princeton, N.J.)
6.2 Strachan, C. (1969) *The Theory of Beta Decay* (Pergamon, New York)
6.3 Wu, C.S., Moszkowski, S.A. (1966) *Beta Decay* (Wiley, New York)

Here, we are particularly interested in the end-point of the energy spectrum giving information about a possible non-vanishing neutrino mass. This is discussed in detail by

6.4 Holzschuh, E. (1992) Rep. Prog. Phys. **55**, 1035
6.5 Holzschuh, E., Fritschi, M., Kündig, W. (1992) Phys. Lett. **B287**, 381

The issue of double beta decay is most important. Geochemical evidence has existed for quite some time but the detailed observation of double beta decay under controlled laboratory conditions is quite recent. We first give a popular text, then some review papers, and a result from recent experiments.

6.6   Moe, M.K., Rosen, S.P. (1989) Scientific American, November, p. 30

6.7   Haxton, W.C. (1983) Comments Nucl. Part. Phys. **11**, 41

6.8   Doi, M., Kotani, T., Takasugi, E. (1985) Prog. Theor. Phys. Suppl. **83**, 1

6.9   Avignone III, F.T., Brodzinski, R.L. (1988) Prog. Part. Nucl. Phys. **21**, 99

6.10 Haxton, W.C., Stephenson Jr., G.J. (1984) Prog. Part. Nucl. Phys. **12**, 409

6.11 Tomoda, T. (1991) Rep. Progr. Phys. **54**, 53

6.12 Beck, M. et al. (1993) Phys. Rev. Lett. **70**, 2853

Neutrino physics has become a very extended domain in physics with topics including neutrino mass, neutrino oscillations, solar neutrino production and detection, etc. We cannot give here a detailed account of the many directions of research but we first refer to some books containing extensive reference lists, to a number of popular accounts, and to some recent review papers and a number of the most basic articles that appeared in the scientific literature.

6.13 Boehm, F., Vogel, P. (1992) *Physics of Massive Neutrinos*, 2nd ed. (Cambridge University Press, Cambridge)

6.14 Bahcall, J.N. (1989) *Neutrino Astrophysics* (Cambridge University Press, Cambridge)

6.15 Winter, K. (ed.) (1991) *Neutrino Physics* (Cambridge University Press, Cambridge)

6.16 Bahcall, J.N., Davis Jr., R., Wolfenstein, L. (1988) Nature, August **334**, p. 487

6.17 Bahcall, J.N. (1994) Beam Line , A Periodical of Particle Physics, Stanford Linear Accelerator Center, Fall, 10

6.18 CERN Courier (1995) June, 13

6.19 Elliott, S.R., Robertson, R.G.H. (1991) Contem. Phys. **32**, No. 4, 251

6.20 Haxton, W.C. (1986) Comments Nucl. Part. Phys. **16**, 95

6.21 Lemonick, M.D. (1996) Time, April 8, p. 46

6.22 Schwarzchild, B. (1986) Physics Today, June, p. 17

6.23 Van Klinken, J. (1995) Ned. Tijdschr. Nat. **11**, 199 (In Dutch)

6.24 Langanke, K., Barnes, L.A. (1996) Adv. Nucl. Phys. **22**, 173

6.25 Oberauer, L., Feilitsch von, F. (1992) Rep. Prog. Phys. **55**, 1093

6.26 Athanassopoulos, C. et al. (1995) Phys. Rev. Lett. **75**, 2650

6.27 Bethe, H.A. (1986) Phys. Rev. Lett. **56**, 1305

6.28 Bethe, H.A. (1989) Phys. Rev. Lett. **63**, 837

6.29 Hampel, W. et al. (1996) Phys. Lett. **B388**, 384 - Most recent GALLEX analyses

6.30 Hill, J.E. (1995) Phys. Rev. Lett. **75**, 2654

6.31 Mikheyev, S.P., Smirnov, A. (1988) Phys. Lett. **B200**, 560

6.32 Wolfenstein, L. (1979) Phys. Rev. **D20**, 2634

We also mention a couple of review papers that concentrate more on observational aspects of neutrino astrophysics, neutrino–nucleus interactions, and theory of supernovae, thereby putting neutrino processes in the context of astrophysics and astronomy.

6.33 Brown, G.E. (ed.) (1988) Phys. Rep. **163**, 1–204

6.34 Koshiba, M. (1992) Phys. Rep. **220**, 229

6.35 Kubodera, K., Nozawa, S. (1994) Int. J. Mod. Phys. **E3**, 101

The implications of free neutron decay for our basic understanding of the standard model are discussed in a recent book presenting neutron properties at length with many references

6.36 Byrne, J. (1994) *Neutrons, Nuclei and Matter* (Institute of Physics, Bristol)

We also mention a recent popular account of the many facets of the neutron and its decay, as well as two articles about experiments that have set error bars as small as possible:

6.37 Gribbin, J. (1993) New Scientist, March, p. 41

6.38 Byrne, J. et al. (1990) Phys. Rev. Lett. **65**, 289

6.39 Stolzenberg, H. et al. (1990) Phys. Rev. Lett. **65**, 3104

The subject of testing fundamental symmetries [parity invariance (P), time reversal (T) , charge conjugation combined with parity (CP), ...] spans a large field of physics, too. We refer to a number of books in order to accommodate the major part of the older literature on this vast subject:

6.40 Roberson, N.R., Gould, C.R., Bowman, C.D. (eds.) (1988) *Tests of Time Reversal Invariance* (World Scientific, Teaneck, NJ)

6.41 Sachs, R.G. (1987) *The Physics of Time Reversal Invariance* (Chicago University Press, Chicago)

A number of review papers concentrating on more recent efforts to test the above symmetries are:

6.42 Adelberger, E.G., Haxton, W.C. (1985) Ann. Rev. Nucl. Sci. **35**, 501

6.43 Henley, E.M. (1969) Ann. Rev. Nucl. Sci. **19**, 367

6.44 Henley, E.M. (1987) Prog. Part. Nucl. Phys. **20**, 387

6.45 Van Klinken, J. (1996) J. Phys. G **22**, 1239

6.46 Wolfenstein, L. (1986) Ann. Rev. Nucl. Part. Sci. **36**, 187

Some interesting, more technical articles are given too:

6.47 Alfimenkov, V.P. et al. (1982) JETP Lett. **35**, 51

6.48 Hayes, A.C. (1996) TTASCC-P-96-2 Preprint
6.49 Müller, A., Harney, H.L. (1992) Phys. Rev. **C45**, 1955
6.50 Severijns, N. et al. (1993) Phys. Rev. Lett. **70**, 4047
6.51 Weidenmüller, H.A. (1991) Nucl. Phys. **A522**, 293c

Finally we give a few more popular accounts:

6.52 Boehm, F. (1983) Comments Nucl. Part. Phys. **11**, 251
6.53 Rosner, J.L. (1987) Comments Nucl. Part. Phys. **17**, 93
6.54 Wolfenstein, L. (1985) Comments Nucl. Part. Phys. **14**, 135

7. The 'Cosmic' Connection: Nuclear Astrophysics and Onwards into the Stars

7.1 Introduction

Recently, a large number of new links between nuclear physics and astrophysics have been made. A totally new discipline has emerged at the intersection of these research fields. It goes under the name of nuclear astrophysics.

Some of the major goals relate to the energy production in stars and are relevant to fields as diverse as basic nuclear physics, astronomy, particle physics, and the study of basic symmetries in physics. A second essential topic that has been pursued for a couple of years albeit in an experimental phase, is that of element synthesis. After the early work of Burbidge et al. it became clear that nuclear physics is the basis for understanding the way all the elements observed in our universe have been 'synthesized'.

In the recent years techniques in nuclear physics made possible the production of radioactive elements and their acceleration up to the energies found in stellar environments. This has enabled the study, under controlled laboratory conditions, of a number of nuclear reactions that are a fundament to element synthesis. The techniques related to radioactive ion beams (RIB) have become a major research topic and will remain so in years to come. Imagine that one can recreate reactions in a controlled way at acclerator facilities – reactions that are responsible for element formation in the universe and thus on earth too.

7.2 Element Synthesis

7.2.1 The Principles and Upwards to ^{56}Fe

In this very short story of element synthesis, which has now been told a few times in manners varying from the strict scientific evolutionary schemes to the more narrative short stories of how the elements as we know today were formed mainly inside stars, we would like to tell the lifestory of a star from the viewpoint of a nuclear physicist.

There are two parts in the road: the first part describes how, through consecutive fusion reactions, elements up to ^{56}Fe have been formed and the second part describes how subsequent neutron capture has led to the formation of many of the heavier elements in the universe.

In the typical burning stages following the big bang, the cooking pot containing gamma radiation, electrons, neutrinos, neutrons, protons, and the lightest elements created in this primordial synthesis constituted most of the matter and radiation of the early universe. Small amounts of deuterium, isotopes of He, and ^{7}Li had been formed at this stage (Fig. 7.1). The earliest stars consisted almost entirely of hydrogen and helium. Most of the heavier elements were produced during stellar evolution and they contribute at most about 2% of the total mass of the universe.

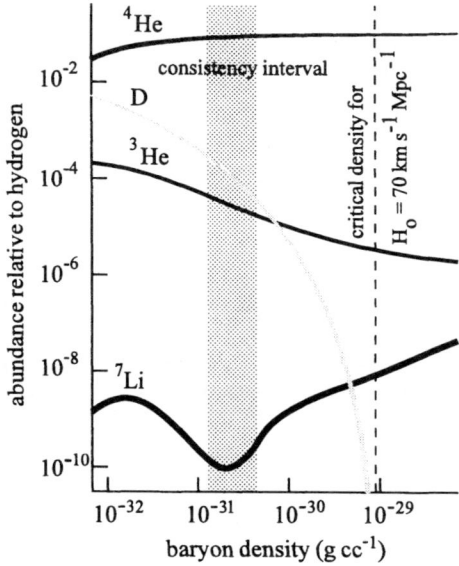

Fig. 7.1. Illustration of the production of light elements in the big bang phase. The primeval abundances of D, ^{3}He, ^{4}He and ^{7}Li can be simultaneously accounted for by a baryon density in the interval $1.5 - 4.5 \times 10^{-31}\,\mathrm{gcm}^{-3}$ (*shaded region*). This density is far below the critical density, shown by the *dashed line*. (Taken from M. Turner (1996) *Physics World* September, Vol.9, No.9. IOP Publishing, with permission)

The elementary burning phase brings the p+p reaction into action near temperatures of $1 - 4 \times 10^7\,\mathrm{K}$ and is at the origin of ^{4}He formation. There are quite a number of variations in the precise nuclear reaction chains but all these chains will inevitably end up with ^{4}He and complete the transformation process from four protons into a single ^{4}He nucleus. As was discussed in Chap. 3, the nuclear binding energy per nucleon increases up to iron, so the above reactions are exothermic. Stars that burn hydrogen into helium are called main sequence stars and about 90% of the lifetime of stars of the appropriate mass is spent in this burning phase.

The p-p process (and also the CNO cycle) gradually transforms the hydrogen in the central region of the star while a He core is forming. During the evolution, the core temperatures continue to rise and reach values of $1.5 - 2.3 \times 10^8\,\mathrm{K}$ which is the threshold temperature for starting the He burning process. Typical nuclear reactions happening at this stage include the fusion of three ^{4}He nuclei into a ^{12}C nucleus and the ^{12}C$(\alpha, \gamma)^{16}$O reaction. The story of the intermediate processes where two alpha particles fuse into a ^{8}Be nucleus with the subsequent capture of a third alpha particle, even within the very short lifetime of ^{8}Be, forms the basis of a "heroic" piece of astrophysics in which Hoyle proposed some of the basic schemes to explain the observational facts.

The above is the typical story associated with the synthesis of the most abundant elements up to mass ^{20}Ne with the exception of ^{14}N, which is formed in the CNO cycle.

At this stage a carbon–oxygen core develops within a core where He gets exhausted in the reaction processes, which is itself surrounded by a H shell. The subsequent burning processes are ignited when the core temperatures, after gravitational contraction, reach the threshold values.

But here, a number of different paths are possible depending on the original mass of the forming star. At present these are not fully known.

In stars with approximately the solar mass, the carbon–oxygen cycle will be the final stage inside the stellar core and most stars of this type will end up as a white dwarf. Stars of a few solar masses up to about eight solar masses will also end their life as a white dwarf, although, the upper mass limit is not so well known. From the nuclear astrophysics point of view, the heavier stars with masses above about eight solar masses are the most interesting ones: Here most of the heavier elements can be formed. In these heavy stars, after further gravitational contraction, carbon burning begins at temperatures of about $6 - 9 \times 10^8$ K. This burning process now starts producing elements such as neon, sodium, magnesium, and small amounts of aluminum. These heavy stars can enter a new steady state of burning. In those conditions, with more neon formed, the next burning stage sets in with neon burning beginning near temperatures of $1.4 - 1.7 \times 10^9$ K. In this burning stage, a large number of high energy photons are produced and photodisintegration starts to become an important factor in the process of producing elements of higher and higher Z value. At this stage, stars become rather complex entities where many nuclear reactions are producing heavier elements, the whole star maintaining a balance between the energy radiated away via photons and neutrinos and the mass of the star tending to contract.

At the end of the carbon and oxygen burning stages, the most abundant nuclei are ^{28}Si and ^{32}S. A star in the appropriate mass region will then start to contract again due to gravitation resulting in the initiation of other reactions. Before these reactions start, the high photon flux will break up (photodisintegrate) a large number of the nuclei already formed. The final outcome of this process is that ^{28}Si is the major species remaining intact.

At temperatures of about 3×10^9 K one reaches a point where photodisintegration of ^{28}Si also becomes possible and also where the Si burning process sets in. In this phase of burning, a very large number of nuclear reactions take place but all have a general tendency of shifting the final nuclei towards the region where nucleons are most strongly bound in the nucleus (i.e., the iron region).

When a star has reached this stage, a variety of nuclear reactions are occurring in the various shells surrounding the rapidly growing iron core. One gets a very specific onion-like burning structure in the star where, like in a big 'cauldron', elements are formed up to the critical phase where nucleons are most strongly bound in a given nucleus and no more energy can be gained by fusing into still heavier elements.

Near to this stage, the mass of a star can no longer be balanced by energy production inside the star and a nova or supernova explosive phase begins. These processes are outside of the domain of this text and we refer to the astrophysics literature for the details. However, we would like to point out that, under these extreme conditions of the stellar evolution, a very precise knowledge of the basic nuclear reactions and nuclear physics are critical for understanding the way in which the 'cosmic' process is evolving.

One of the obvious questions now is how the heavy elements up to the uranium region could have been formed. The answer is that a sequence of neutron capture processes and subsequent beta decay processes in competition, known in astrophysical language as the slow (s process) and rapid (r process), make possible the long route to the formation of the heavier nuclear species. These processes form a big chapter in the study of astrophysics. They represent the fine-tuned cooperative action of nuclear physics processes within an extreme environment of stellar burning and have been described in great detail. This description forms one of the great epics of our basic physical knowledge, going all the way back to the early universe.

7.2.2 Slow and Rapid Neutron Capture

The formation of heavy elements far beyond ^{56}Fe, the seed nuclei, is the result of a unique set of nuclear processes in which fusion can no longer play any role due to the endothermic nature of the fusion process beyond ^{56}Fe. The heavy elements are formed instead through a chain of neutron capture reactions with subsequent beta decay. The difference between slow (s process) and rapid (r process) capture reactions will be discussed in some detail because here, the nuclear physics plays an important role in our understanding of element formation and in explaining, e.g., the abundance of elements in our solar system.

The neutron capture primarily forms elements starting from the Fe seed nuclei. The source of the rather low energy neutrons is reactions like (α, n) on ^{22}Ne and ^{13}C where the alphas are highly abundant during He burning in the stars. The small neutron abundance then causes a particular path to be formed in which the half-life for neutron capture is much longer than the beta decay half-life. So, this particular condition implies a path that lies mainly along the valley of beta stability because the capture process follows a time scale dictated by the various beta decay processes (see Fig. 7.2 for an illustration of this process). One can follow this process in great detail starting at ^{56}Fe throughout the mass table. One thus finally arrives at the element ^{210}Bi which is unstable with respect to α-decay and one observes a cycling back into ^{206}Hg, thereby ending the s process.

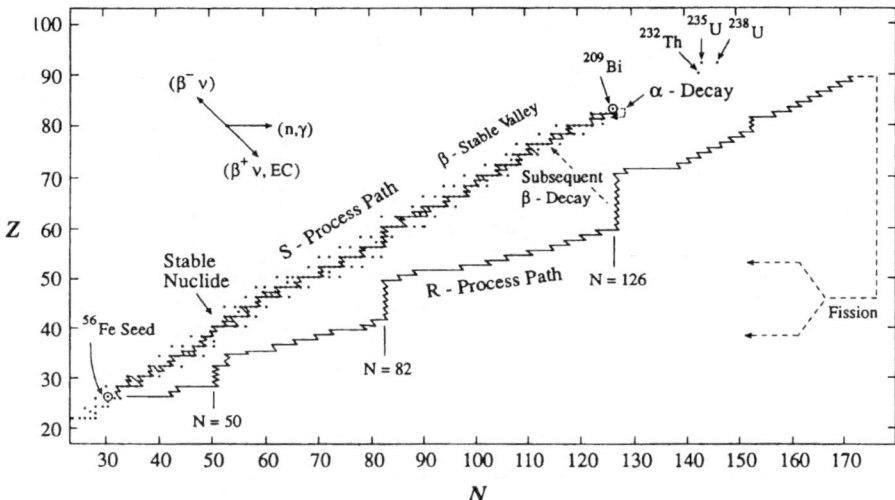

Fig. 7.2. A schematic representation of the mass table showing in particular the r process, in which heavy elements are formed through a series of rapid neutron capture and beta-decay processes far from the region of beta-stability, and the s process path. (Taken from R.P. Wang et al. (1992) AAPPSB Bulletin, June, Vol. 2, No. 2, with permission)

One of the prerequisites for the whole process to take place steadily is a large enough neutron density. These conditions are fulfilled in particular during the He burning stage in stars. When, at a later stage, these elements get ejected into the interstellar medium through novae and supernovae events, the basic material in the universe is constantly being enriched. There is clear-cut evidence for sharp peaks in the abundance curve near neutron closed shell configurations which can only result from the existence of such a slow neutron capture process (Fig. 7.3).

On inspecting the abundance curve in some detail, one finds evidence for a number of additional broad peaks at around 8–12 mass units prior to the s-process abundance peaks. This indicates that other capture processes where neutrons are involved have to take place in order to produce those elements as well being responsible for the production of elements beyond ^{209}Bi. This is called the rapid or r process.

In order for this process to work effectively, one needs a very large neutron flux, with densities typically of the order of 10^{20} cm^{-3}. Here, the process is such that a given starting nucleus is capturing a large number of neutrons and

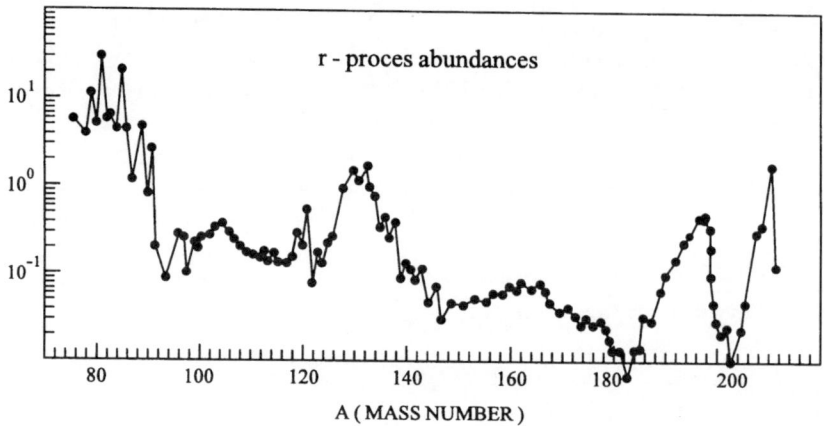

Fig. 7.3. The r-process abundances as a function of the mass number A. Sharp and pronounced peaks at $A = 80, 130$ and 195 are clearly visible. (Adapted from insert in Fig. 15. in *The Isospin Laboratory: Research Opportunities with Radiactive Nuclear Beams*, Report LALP 91–51, with permission)

so proceeds along an isotopic chain. This long sequence only stops if (i) an isotope is formed with a very short lifetime for beta decay, or (ii) the (γ, n) process occurs much faster than the neutron capture. These conditions, as is seen by inspecting known mass tables and studying properties of highly unstable nuclei, are reached almost 20–30 mass units beyond the stable isotopes for a given element. An r-process path then evolves along the neutron drip line region and can be derived under the waiting point approximation which takes into account that the time scale for neutron capture is much shorter than that for the beta decay processes. So, before beta decay starts, an equilibrium in abundances builds up.

Astrophysicists, generally, believe that the r-process nucleosynthesis sites are situated in the inner and neutron-rich ejected layers of stars after a supernovae explosion. Depending on the neutron density at the various places, different elements will form the starting point and so different r paths will result. In this process almost all elements between the line of stability and the neutron drip line will take part in the r-process nucleosynthesis and this amounts to about 5000 nuclei. In this process, reaching far out beyond known nuclei, theoretical mass formulas that predict nuclear masses are greatly needed. There are still many uncertainties in extrapolating our present-day knowledge of nuclear physics (Chap. 3) towards the region of the drip line where weakly bound nuclei are obtained. This field of nuclear physics is itself one of the major topics of research at present, but contains in addition much potential for astrophysics purposes. Besides theoretical nuclear physics studies, experimentalists are trying to approach as close as possible to this drip line region. The known stable nuclei are too far away from the drip line region but, with the advent of new facilities that allow the acceleration of radioactive ion beams and yield the appropriate target–radioactive element (beam) combination, one can explore new regions of nuclear matter and gain a better understanding of these extreme conditions.

At some stage, the neutron density and temperature in the stellar environment will drop and so the conditions for r capture cease (or "freeze out"). The final result is that the r-capture elements decay back towards the valley of beta stability leading to the formation of, e.g., the actinide elements which have very long beta decay half-lives.

It has become clear that, in the course of the element formation, the initial nucleus through which a given path passes is unstable (radioactive) in most cases. This insight yields a better understanding of the CNO cycle, the s-, and r-processes. With the advent of new experimental techniques in creating intense beams of radioactive elements, a number of critical stellar reactions can now be carried out under controlled laboratory conditions. This is an important motivation for constructing a variety of radioactive ion beams which can explore and study the 'nuclear ashes' formed in the stellar interior. The last few years have therefore seen something of an 'explosion', with a large number of proposals for creating (i) radioactive ion beams with well-defined astrophysics purposes and (ii) facilities that will provide astrophysicists and nuclear physicists with a large spectrum of radioactive elements with energies up to and beyond the Coulomb barrier energy. These features will be discussed in the next two sections.

7.3 Why Radioactive Ion Beams Are Needed and How to Produce Them

7.3.1 Why Do We Need Radioactive Ion Beams?

As discussed in the previous two subsections, the production of the many elements that are observed in our present-day universe is the result of burning processes inside stars, on top of which a very large variety of nuclear reaction processes act to synthesize the heavier elements and the proton rich nuclei. Because the temperatures are high enough to create charged particles with energies that approach the Coulomb barrier energy, the nuclear burning occurs on a timescale of seconds.

A quantitative understanding of the observed abundances requires a knowledge of cross-sections such as (p, γ) and (α, γ), and of the many other types of reactions involved. For many elements, half-lives are of the order of a day or longer. In such cases, radioactive targets can be constructed and the subsequent reactions studied. In the majority of the most interesting cases, however, half-lives are too short (of the order of a few seconds or minutes) so that beams of these radioactive elements have to be made and accelerated up to the appropriate energies in order to induce the critical nuclear reactions determining the precise paths in the (N, Z) mass plane in the synthesis of the elements. The two major production methods will be discussed in some detail and, in the next section, applied to a number of reactions in the CNO cycle of light element formation.

Charged-particle induced reactions are very important for the study of light elements up to the Fe region. In this domain, the theoretical calculations using Hauser–Feshbach methods are far too inaccurate because the presence of certain resonances can dominate a given reaction at the appropriate energy. In a number of cases, inverse kinematics is preferable, but in those cases too, restrictions are present because the levels to be reached in the original or inverse reactions can often be quite different.

Another class of reactions that carry essential information about nuclear processes inside stars are the various capture reactions like (p, γ) and (α, γ) for which the inverse reactions can be studied using Coulomb break-up in the field of a heavy nucleus.

In most cases, direct measurement of the various cross-sections, using the appropriate RIB, is by far the best approach but it may often be a technically difficult task to produce the appropriate beams and accelerate them to the energies needed.

Radioactive ion beams are particularly useful at energies up to 1 MeV/A with high intensities allowing the measurement of cross-sections as discussed before. On the higher energy side (10–20 MeV/A) one can determine reaction cross-sections in another, otherwise inaccessible, energy window, albeit in an indirect way.

7.3.2 How Can Radioactive Ion Beams Be Created?

We shall now discuss the various ways of producing radioactive ion beams (RIB) and a number of the latest developments in this field.

Exotic beam facilities can be characterized by the production mechanism of the radioactive elements, which will subsequently be accelerated up to the appropriate energy. The isotope separator on line (ISOL) technique and the projectile fragmentation method are the two basic approaches. They are depicted schematically in Fig. 7.4. The precise outline for a typical example of both (Louvain-la-Neuve for the former and GANIL for the latter) will be discussed in more detail in a technical (Box XI) on RIB facilities.

In the ISOL method, radioactive isotopes far away from stability are produced by bombarding a thick target with a high-intensity beam of protons, alpha particles, or light ions. The radioactive nuclei are separated out of the target in the form of atoms or even molecules. Then they are ionized within a specific ion source (this is in most cases one of the most difficult technical steps along the chain) and subsequently accelerated up to the required energy in a second accelerator. These radioactive ion beams, also called secondary beams, are then themselves directed towards a target in order to study the particular nuclear reaction of interest.

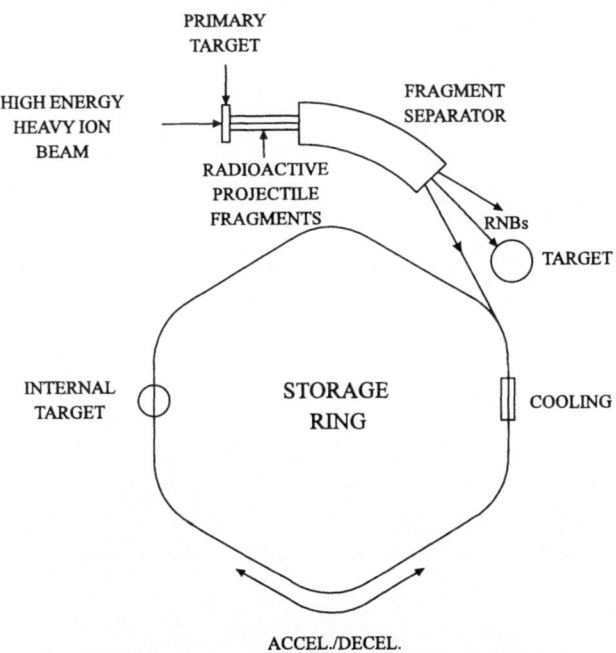

Fig. 7.4. Illustration of the ISOL (*upper part*) and projectile fragmentation (*lower part*) methods for producing high-intensity radioactive ion beams. (Taken from *Proposal for Physics with Exotic Beams at the Holifield Heavy-Ion Research Facility* ©1991, eds. J.D. Garrett, K.D. Olsen)

Table 7.1. Existing and proposed ISOL exotic beam facilities. (Taken from *Proposal for Physics with Exotic Beams at the Holifield Heavy-Ion Research Facility* ©1991, eds. J.D. Garrett, K.D. Olsen)

Facility	Driver Accelerator	RIB Accelerator	Mass Range	Maximum Energy*	Status
Louvain-la-Neuve	$K = 30\text{H}^-$ cyclotron 30-MeV p	$K = 120$ cyclotron	13	0.65	Achieved
ARENAS 3 Louvain-la-Neuve	$K = 120$ cyclotron 80-MeV p	45 MV SC linac	≤ 30	1.5	Proposed
HHIRF ORNL	$K = 100$ cyclotron 60-MeV p	25 MV tandem	< 80	5–13	Operational
EB88 LBL	$K = 30\text{H}^-$ cyclotron 30-MeV p	$K = 140$ cyclotron	< 30	$140\,(Z/A)^2$	Discussed
ISAC TRIUMF	$K = 500\text{H}^-$ cyclotron 500-MeV p	Linac	< 60	1.5	TISOL exists Proposed
PRIMA CERN	1.0-GeV p synchrotron	Linac	≤ 27	1.4	ISOLDE-3 exists Proposed
JHP KEK	1.0-GeV p linac	Linac	< 60	6.5	Proposed
Rutherford Lab	800-MeV p synchrotron ISIS	Linac or synchrotron (into ISIS)	≤ 80	6.5 45–120	Under study
ISL*	–	–	< 220	10	Under study

* in Units: MeV per nucleon, or MeV/A

There are of course many variants of this scheme for producing RIB, several of which are at present still on the drawing board or under construction. Variations exist, for example, in the choice of the light elements to accelerate into the heavy target in the initial phase. Here, in the primary beams, one can use protons of rather low energy (20–50 MeV) or very high energy (1 GeV at CERN) proton beams, or one can even use heavy ions. There are also several ways of extracting the nuclei produced and of ionizing them (we

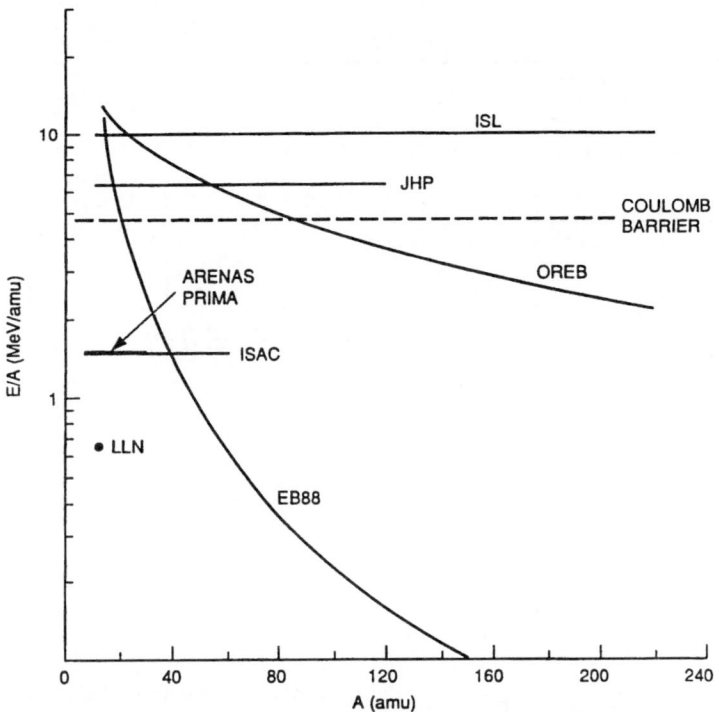

Fig. 7.5. Comparison of the maximum energy per nucleon (E/A) as a function of the nuclear mass number A for various existing and proposed exotic and radioactive ion beam facilities, based on the ISOL technique. The Coulomb barrier height is also given for comparison. (Taken from *Proposal for Physics with Exotic Beams at the Holifield Heavy-Ion Research Facility* ©1991, eds. J.D. Garrett, K.D. Olsen)

refer to specialized literature on this point) and, subsequently, of choosing a second accelerator to give the radioactive ions the desired energy.

At present, there is only one such facility actively working at Louvain-la-Neuve with the HHIRF just becoming operational. This will be discussed and outlined in more detail in a technical box (Box XI). In brief, it consists of a 30 MeV, 500 µA proton cyclotron as a first accelerator and an electron cyclotron resonance (ECR) ion source followed by a $K = 110$ MeV second cyclotron accelerator. Here the K value is connected to the maximum energy per nucleon delivered by the cyclotron, W^{max}, through the relation

$$W^{\mathrm{max}} = K\,(q/A)^2 \tag{7.1}$$

with q the ion charge and A the atomic mass number. For a proton, one obtains $W^{\mathrm{max}} = K$. Running this combination in the optimal mode, one can produce RIB with energies of 0.5 MeV/A and $10^8 - 10^9$ particles per second. Quite a number of other projects (see Table 7.1) will be based on this technique: in Europe (at GANIL, Catania, ISOLDE CERN, PIAFE at Grenoble)

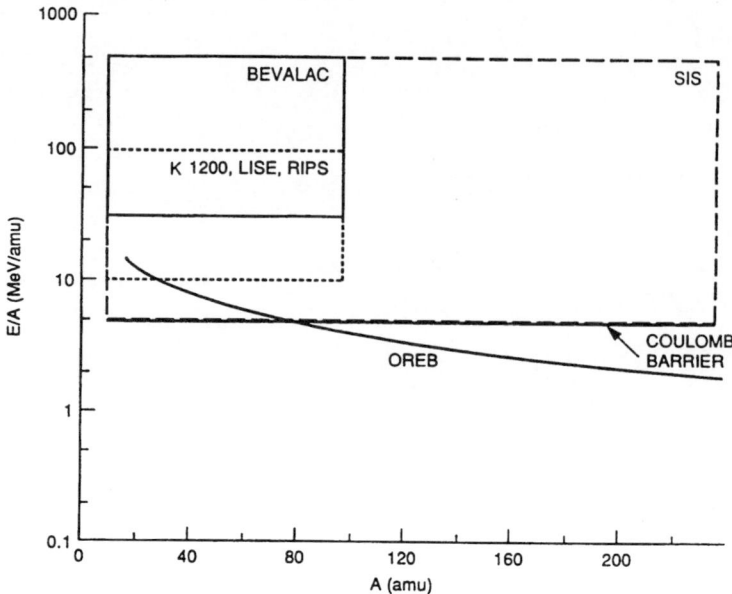

Fig. 7.6. Comparison of the energy per nucleon (E/A) as a function of the nuclear mass number A for various existing projectile fragmentation exotic beam facilities together with the maximum E/A value for the proposed extension of the HHIRF. The Coulomb barrier height is also shown for reference. (Taken from *Proposal for Physics with Exotic Beams at the Holifield Heavy-Ion Research Facility* ©1991, eds. J.D. Garrett, K.D. Olsen)

and in the USA (Oak Ridge), in Japan (ISN, Tokyo), and in Canada (TRI-UMF).

A second, important method relies on a different mechanism of nuclear fragmentation. Here, a high-energy beam of stable heavy ions is accelerated onto a target of light nuclei with the subsequent fragmentation of the projectile into many nuclei, of which a large number are radioactive and which are emitted almost exclusively in the forward direction. The desired RIB with energies almost of the order of the primary beam are separated from the primary beam and from the other fragments (the details can be found in many technical reports and articles) and are directed towards a second target for experiments to be performed.

Here again, as the ISOL technique, there are a number of variants that already exist or are being planned for various energy of the primary beams, various separation methods, and, sometimes, with the addition of an extra ring that can accumulate radioactive ions and cool them for specific purposes. Examples of this method are operating at GANIL (Caen), RIKEN (Japan), MSU (Michigan, USA) with a number being designed and/or under construction.

The maximum energy reached (per nucleon, expressed as E/A) is illustrated for both the ISOL technique (Fig. 7.5) and the projectile fragmentation method (Fig. 7.6).

Box XI

A Short "Tour" of Radioactive Beam Facilities

Unique possibilities have opened up with the recent and forthcoming radioactive ion beam facilities. They will allow the study of a large domain of physics, extending across various borderlines that previously separated nuclear physics from, e.g., astrophysics, cosmology, particle physics, or solid state physics.

In this technical box we discuss in some detail two typical existing facilities concentrating on the prospects for carrying out experimental programs related to nuclear astrophysics. These are: the project at Louvain-la-Neuve, a typical ISOL type for the production of secondary beams and GANIL, the "Grand Accelerateur National d'Ions Lourds" which produces higher energy radioactive ions beams using projectile fragmentation.

Finally, we survey the existing projects and their stage of development.

The approach used at Louvain-la-Neuve, as outlined before, comprises two cyclotrons, coupled on line via an intermediate ion source of the electron cyclotron resonance (ECR) type. It is outlined in Fig. XI.1.

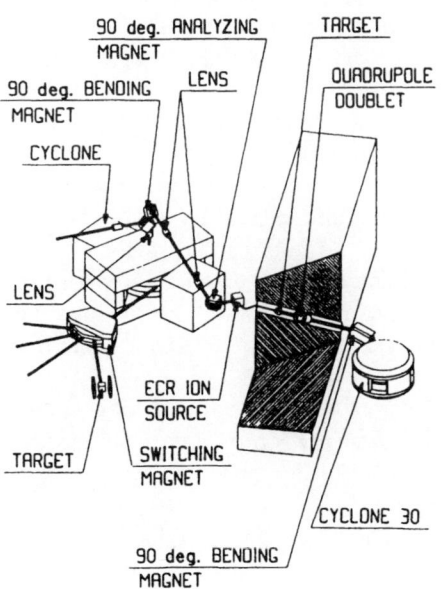

Fig. XI.1. Schematic drawing of the radioactive ion beam facility at Louvain-la-Neuve. A two cyclotron setup is used here. (Taken from Darquennes et al. (1990) Phys. Rev. **C42**, R804. American Physical Society, with permission)

The first CYCLONE 30 accelerator is used to produce large numbers of radioactive nuclei that are extracted from the first target, made into atomic or molecular form, and ionized in the ECR ion source. Following a transport phase, the radioactive ions are then injected into the second stage cyclotron. This accelerator delivers the RIB with the desired energy to study nuclear reactions at the end station.

Some of the main advantages of such a two-accelerator set up are: (i) a broad spectrum of radioactive ion beams can be produced over an energy interval of 0.2 to about 2 MeV/A, energies that are particularly suited for the study of reactions of astrophysical importance. Here, the first radioactive ion beams of ^{13}N in a 1^+ charge state have been produced with an energy of 0.65 MeV/A and an intensity of 1.5×10^8 particles/s. This was made possible with the 30 MeV, 100 μA proton beam produced at the CYCLONE 30 cyclotron, which was directed at a target located in a large concrete wall used to separate the vaults of the two cyclotrons. Large amounts of ^{13}N were produced there through the reaction ^{13}C(p, n)^{13}N. The ^{13}N was then extracted from the original ^{13}C target in the form of N_2 molecules (^{13}N^{14}N). These molecules were transferred to an ECR ionization source, especially designed to optimize the production of the 1^+ charge state in the atomic ions. These particular ions were subsequently extracted from the ECR source, analyzed according to mass, transported, and injected into the second cyclotron CYCLONE. The latter cyclotron has been modified in particular to optimize the acceleration at the lower energy side so as to produce accelerated radioactive ions in the energy region of 0.65 MeV/A. These ions are particularly well suited for studying reactions of astrophysical significance. The accelerated ^{13}N 1^+ ions were extracted from CYCLONE, directed through a switching magnet towards a thick stopper plate and the β^+ activity deposited there was measured. It was confirmed that the radioactivity was indeed due to ^{13}N by controlling the decay half-life.

The first physics experiment carried out with this beam was the study of the ^{13}N(p,γ)^{14}O reaction which is crucial to understanding parts of the hot CNO cycle. The aim was to study the direct gamma capture width through the direct proton capture into the dominant resonance at 0.445 corresponding to the above stellar temperature scenario. A number of other dedicated radioactive ion beams have been developed since and are included in Table 7.2 of the main text.

The GANIL accelerator facility at Caen in France has a long-standing tradition of accelerating a large spectrum of heavy-ion beams. Much has been learnt since 1979 about projectile fragmentation for producing and studying reactions induced by radioactive ion beams.

At present, GANIL produces high intensity beams for ions from ^{12}C up to ^{238}U at energies of 96–24 MeV/A. Great efforts have been made to improve the beam intensity with the aim of later producing radioactive ion beams. To reach this goal, the primary heavy-ion beam accelerated by the

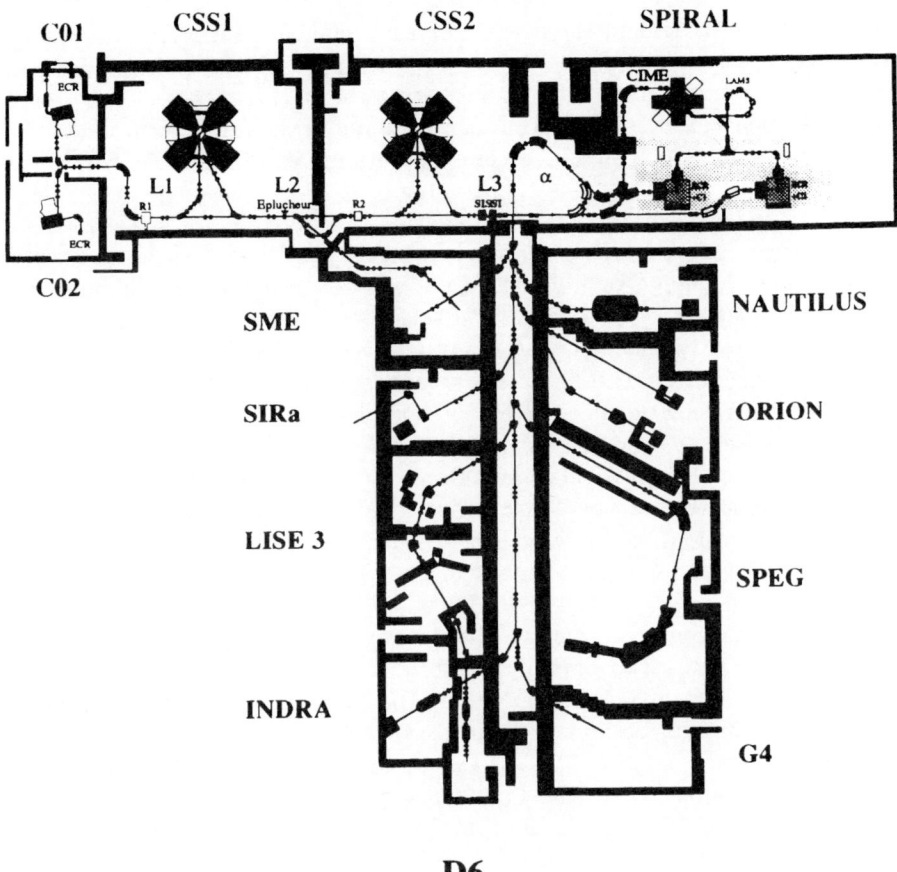

Fig. XI.2. Layout of the future installation SPIRAL at GANIL. The primary beam of the CSS2 cyclotron can be directed either to the present experimental area or to SPIRAL. After production, ionization and magnetic separation, the secondary beam is injected into the CIME cyclotron. The accelerator beam is then sent to the experimental hall. (Taken from A.C.C. Villari et al. (1995) Nucl. Phys. **A588**, 267c and M. Lieuvin, Spiral Project Leader. Elsevier Science, NL, with kind permission)

GANIL cyclotrons (see Fig. XI.2) will bombard a production target in a well-shielded underground area. The ten-fold increase in intensity, available at present (1996), is at the same time beneficial for the high-energy radioactive ion beams prepared by projectile fragmentation. The radioactive ions thus produced will be at high temperature (about 2300 K) and will pass into a ECR ionization source. After extraction from this source, the low-energy RIB will be selected by a relatively low resolution separator ($\Delta m/m = 4 \times 10^{-3}$) and injected into a new cyclotron CIME. The design and construction of this post-accelerating cyclotron (a $K = 265$ cyclotron) is the result of a collaboration between GANIL and the Institut de Physique Nucleaire (IPN) at Orsay. After acceleration, with final energies between 1.8 and 25 MeV/A, a special magnetic selection will be carried out before the RIB is directed into one of the experimental areas. The details can be found in the full description of the project SPIRAL which should become operational in 1998 with experiments planned for the end of 1998 and beginning of 1999.

Thus, at GANIL, the unique combination of producing, in the same place, a spectrum of ISOL radioactive ion beams, as just described, and in flight production of high-energy radioactive ion beams through projectile fragmentation will clearly form the basis for a very broad research program.

Besides these two facilities, several upgrades of existing facilities and new projects are coming on line. Extensive and comparative studies have been made, concentrating on the specific and complementary possibilities of these various facilities. A very interesting publication is the "Isospin Laboratory" newsletter which brings regular news from the various laboratories producing RIB; this newsletter can be consulted to follow developments in the coming years. In conclusion, we draw attention to the important investments, both in R&D and in fundamental physics, that are being made at the following institutions: Oak-Ridge (HRIBF – Holifield Radioactive Ion Beam Facility), NSCL (National Superconducting Laboratory at Michigan State University), ATLAS at Argonne National Laboratory, a radioactive ion beam project ISAC at TRIUMF in Canada, upgrading the Notre-Dame/Michigan existing facilities. In Europe, besides Louvain-la-Neuve and the SPIRAL project at GANIL, there are projects at ISOLDE CERN (the REX-ISOLDE project), PIAFE in Grenoble where the high-flux reactor at the ILL is being linked with the cyclotron facility at the ISN. In Japan, the RIKEN facility which is already operating, has major plans to construct a versatile and large-scale RIB facility.

7.4 Studying the Nuclear Ashes in the Laboratory

With the present means of accelerating radioactive ion beams, one can study a number of key reactions in the long chain of element synthesis, from protons right through to the heaviest elements known in the universe today.

We shall discuss some of the reactions that have been studied in detail by using low-energy light radioactive ion beams, concentrating in particular on the reaction ^{13}N(p, γ)^{14}O, a key reaction in understanding the CNO cycle.

Many stars, similar to our sun, are emitting the energy produced in the burning of four protons into a ^{4}He nucleus via a chain called the CNO (carbon–nitrogen–oxygen) cycle, first discovered by H. Bethe. For this discovery he later received the 1967 Nobel prize in physics. This cycle is shown in

Fig. 7.7 and illustrates the catalytic action of the ^{12}C in the transformation of four protons into a single ^{4}He nucleus with the simultaneous liberation of a large amount of energy. There is an interesting point in the study of this chain: At a large enough temperature, in what is called the 'hot' variant of the original (now called 'cold') CNO cycle, the radioactive element formed, ^{13}N, may capture a proton and so form ^{14}O which then can decay back into the stable element ^{14}N. This variant is particularly important because it may lead to a 'break-out' from the CNO cycle leading to the formation of heavier elements. In order to draw quantitative conclusions one has to know in much more detail the cross-section for proton capture in ^{13}N with the subsequent formation of ^{14}O. This proton-induced reaction is dominated by a resonance at 0.445 MeV center-of-mass energy in the ^{13}N+p system, which corresponds to the first excited state in ^{14}O at 5.17 MeV.

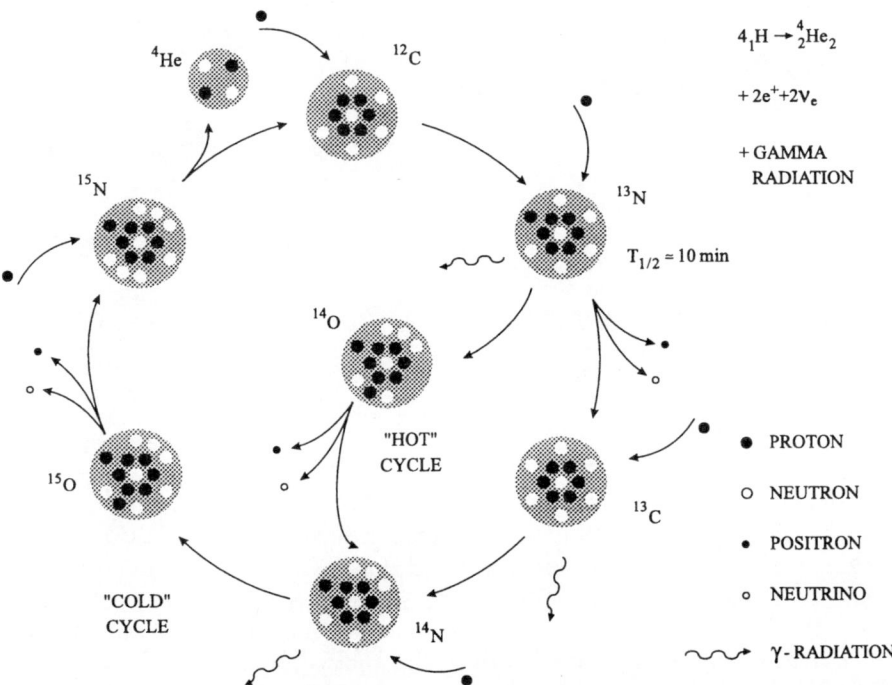

Fig. 7.7. The various reaction steps characterizing the 'cold' and the 'hot' CNO cycles, starting from ^{12}C and producing ^{4}He from four protons. (Taken from M. Huyse, P. van Duppen (1991) *Natuur en Techniek* October, with permission)

Concluding, this reaction depends critically on the possibilities for producing the radioactive element ^{13}N in sufficiently large numbers. This beam must then be accelerated up to the appropriate energy to induce the ^{13}N+p reaction. In a typical star, ^{13}N is formed within a time scale that is large on the astronomical level. Under laboratory conditions, we can create the appropriate isotope by bombarding ^{13}C with protons. One has to remember that natural carbon consists to 98.9% of ^{12}C and only 1.1% of the isotope ^{13}C. So, in the first step of acceleration, the first cyclotron accelerates protons to create the necessary ^{13}N nuclei which have to be separated out of the target, ionized, transported to the cyclotron, and then themselves accelerated in order to induce the ^{13}N+p reaction in inverse kinematics (see also Fig. 7.8). In the first stages of producing the appropriate RIB of ^{13}N, a huge contamination of ^{13}C was present because, after the reaction on the first target, in addition to separating out the radioactive ^{13}N, the element ^{13}C with almost the same mass is very difficult to distinguish and also gets ionized and brought into the second accelerator phase. After a large number of technical improvements, a sufficiently intense beam was produced, allowing the reaction ^{13}N(p, γ)^{14}O to be studied.

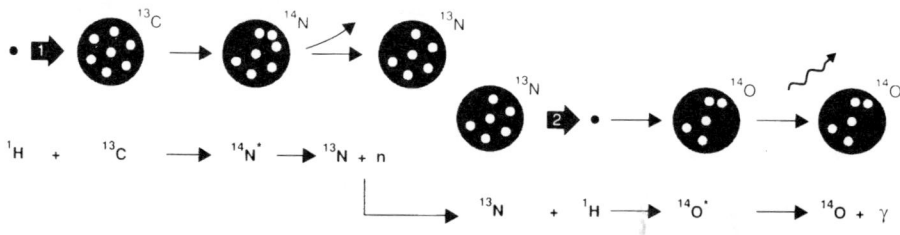

Fig. 7.8. The two steps, drawn schematically as separate reaction channels, needed to produce ^{14}O. In the first phase, the p + ^{13}C → ^{14}N* → ^{13}N + n reaction is performed using CYCLONE-30. Subsequently, ^{13}N, as a radioactive ion, is accelerated in the second cyclotron CYCLONE and induces the ^{13}N + p → ^{14}O* → ^{14}O + γ reaction. (Taken from M. Huyse, P. van Duppen (1991) *Natuur en Techniek* October, with permission)

Three different types of measurements have been performed in this case: (i) the direct determination of the proton-induced reaction cross-section integrated over the 0.445 resonance, which has yielded the radiative width of this particular resonance; (ii) the determination of the spectroscopic factor for the ^{13}N(d, n)^{14}O (ground-state) transition; and (iii) a direct measurement of the energy and width of this resonance through the study of the ^{13}N +p elastic scattering in the resonance region. All these data have allowed a number of important astrophysical quantities to be calculated and have also confirmed experimentally, under laboratory constraints, the importance of the ^{13}N(p, γ)^{14}O reaction in nova processes.

In Table 7.2, we indicate the characteristics of various accelerated RIB at the facility in Louvain-la-Neuve.

Another interesting capture reaction is the ^{19}Ne(p, γ)^{20}Na reaction which could lead to escape from the hot CNO cycle and is situated in the chain

$$^{15}\text{O}(\alpha, \gamma)^{19}\text{Ne}(p, \gamma)^{20}\text{Na}(\beta^+)^{20}\text{Ne} , \qquad (7.2)$$

Table 7.2. Characteristics of the accelerated radioactive ion beams available at the ARENAS 3 facility. (Taken from M. Huyse et al. (1995) Nucl. Phys. **A588**, 313c. Elsevier Science, NL, with kind permission)

Isotope	$T_{1/2}$ (s)	Production target	E_{max} (MeV)	Intensity (pps)
$^6\text{He}^+$	0.8	LiF	18	3×10^6
$^{11}\text{C}^+$	1200	BN	10	1×10^7
$^{13}\text{N}^+$	600	^{13}C	8.5	2×10^8
$^{13}\text{N}^{2+}$	600	^{13}C	34	1×10^8
$^{18}\text{Ne}^{3+}$	1.7	LiF	55	4×10^5
$^{19}\text{Ne}^{2+}$	17	LiF	23	5×10^8
$^{19}\text{Ne}^{4+}$	17	LiF	93	2×10^8
$^{35}\text{Ar}^{5+}$	1.8	NaCl	79	1×10^5

and may so be the starting point for rp (rapid proton capture reaction) processes, in which, through succesive proton capture reactions starting from elements in the mass $A = 20$ region, nuclei with masses up to $A = 60 - 75$ could have been formed in the proton rich sector of the mass table. This reaction too has been studied in a direct way using the radioactive ion beam of ^{19}Ne made at the Louvain-la-Neuve facility.

Interesting information for astrophysics has been obtained with the very low energy RIB at on-line isotope separators producing elements with 50–100 keV energies. Examples are the study of waiting point nuclei in the r process, for instance, ^{79}Cu and ^{130}Cd at ISOLDE CERN. More recently, the beta decay of ^{16}N to unbound levels in ^{16}O has been studied in detail by research groups at TRIUMF and Yale and is giving information on one of the most critical and astrophysically imporant quantities, namely the cross-section of the $^{12}\text{C}(\alpha, \gamma)^{16}\text{O}$ reaction.

The use of higher energy RIB of various elements is important for nuclear astrophysics too because it allows one to obtain information on cross-sections in key reactions mainly through Coulomb dissociation. This method relies on the intimate relation between a given nuclear reaction and its time-reversed reaction (principle of detailed balance) in order to determine cross-sections for radiative capture of a given particle and the inverse photodisintegration reaction cross-section $X(\gamma, \text{particle})Y$. In these reactions one does not use real photons but the virtual photon spectrum which is generated when a radioactive ion beam passes through the field of a heavy nucleus like ^{208}Pb. This method has the advantage over the direct capture mechanisms of the particle in the very fact that cross sections are on average a factor 1000 more important in the dissociation experiments. Thus one obtains much better counting statistics and/or needs lower intensity RIB. There are, of course, limitations too: for example, one cannot ensure that the Coulomb excitation is the sole reaction process. Nuclear interactions may also enter and complicate the final reaction through unwanted interferences with the pure Coulomb amplitude. Various other shortcomings are discussed in recent articles and reviews of the various accelerating schemes for obtaining effective and selective RIB (see the reference list at the end of this chapter [7.27]–[7.40]).

Experiments have been carried out using this method at RIKEN and at GANIL, also studying the $^{13}\mathrm{N}(\mathrm{p}, \gamma)^{14}\mathrm{O}$ reaction, but here using the photodissociation method. Results are in good agreement with the direct method used at Louvain-la-Neuve. A number of the reactions that are accessible almost exclusively through this method are of great astrophysial interest. RIKEN and GANIL will thus develop further research programs concentrating in particular on the light nuclei that are critical in understanding the proton burning phase in stars.

7.5 Conclusion

In this chapter we have pointed out the various cross connections between astrophysics and nuclear physics. We have seen that efforts to built a set of full-scale facilities able to create radioactive ion beams with a large range of (Z, A) and a broad range of energies, may solve a number of crucial questions concerning element synthesis in stars. The chains of element formation that have long been a unique characteristic of the heavens have now moved into the nuclear physics laboratories. Beams of selected elements have already been able to confirm our ideas about the CNO cycle of hydrogen burning. More extensive efforts towards the end of the present decade and in the new millenium will undoubtedly shed new "light" on the production and abundance of the elements that form our present universe.

7.6 Further Reading

First of all, we present a few text books that concentrate on nucleosynthesis in the context of more general nuclear astrophysics problems. These books give a very general and extensive coverage of the major issues.

7.1 Arnett, W.D., Truran, J.W. (eds.) (1985) *Nucleosynthesis: Challenges and New Developments* (University of Chicago Press, Chicago)

7.2 Clayton, D.C. (1983) *Principles of Stellar Evolution and Nucleosynthesis* (University of Chicago Press, Chicago)

7.3 Kippenhahn, R., Weigert, A. (1990) *Stellar Structure and Evolution* (Springer, Berlin Heidelberg)

7.4 Phillips, A.C. (1994) *The Physics of Stars* (Wiley, New York)

7.5 Rolfs, C.E., Rodney, W.S. (1988) *Cauldrons in the Cosmos* (University of Chicago Press, Chicago)

We refer to the basic article by F.M. Burbidge et al. on element synthesis as well as the Nobel Prize account by W. Fowler, a most readable review article:

7.6 Burbidge, F.M., Burbidge, G.R., Fowler, W.A., Hoyle, F. (1957) Rev. Mod. Phys. **29**, 547
7.7 Fowler, W.A. (1984) Rev. Mod. Phys. **56**, 149

In relation to the various processes that account for both the standard synthesis (up to ^{56}Fe) and the production of medium-heavy and heavy nuclei, we first present a number of more popular accounts:

7.8 Astroparticle Physics – Special Issue (1996) Physics World, September, Vol. 9, No. 9, pp. 29–56
7.9 Bethe, H.A., Brown, G.E. (1985) Scientific American, April
7.10 Bignami, G.F. (1987) Nature, **325**, 302
7.11 Levi, B.G. (1993) Physics Today, July, 23
7.12 Mathews, G.J., Cowan, J.J. (1990) Nature, **345**, 491
7.13 Pagel, B.E.J. (1991) Nature, **354**, 267
7.14 Pasachof, J.M., Fowler, W.A. (1974) Scientific American, May
7.15 Viola, V.E., Mathews, G.J. (1987) Scientific American, May
7.16 Wang, R.-P., Lu, N.-Y., Feng, D.-H, Thielemann, F.-K. (1992) AAPPSB (Assoc. of Asia Pacific Phys. Soc. Bulletin) Vol.2, No.2, 2

A number of more technical review articles and more detailed accounts of the above processes:

7.17 Bethe, H.A. (1990) Rev. Mod. Phys. **62**, 4
7.18 Chen, B. et al. (1995) Phys. Lett. **B355**, 37
7.19 Copi, C.J., Schramm, D.N., Turner, M.S. (1995) Phys. Rev. Lett. **75**, 3981
7.20 Cowan, J.J., Thielemann, F.-K., Truran, J.W. (1991) Phys. Rep. **267**, 208
7.21 Han, X.-L., Wu, C.-L., Feng, D.-H., Guidry, M.W. (1992) Phys. Rev. **C45**, 1127
7.22 Hata, R. et al. (1995) Phys. Rev. Lett. **75**, 3977
7.23 Kappeler, F., Beer, H., Wisshak, K. (1989) Rep. Prog. Phys. **52**, 945
7.24 Schramm, D.N. (1995) Nucl. Phys. **A588**, 277c
7.25 Voss, F., Wisshak, K., Guber, K., Käppeler, Reffo, G. (1994) Phys. Rev. **C50**, 2582
7.26 Wang, R.-P., Thielemann, F.-K., Feng, D.-H., Wu, C.-L. (1992) Phys. Lett. **B284**, 196

Recently, new experimental techniques have made it possible to accelerate radioactive ion beams (RIB) thus allowing hitherto impossible reactions starting from unstable nuclei. The development of a variety of RIB has, in particular, allowed the nuclear reactions responsible for the synthesis of the light elements (CNO cycle) to be performed under controlled laboratory conditions. The literature about this field is vast and there are also a number of reports putting the case for new RIB at various existing and new facilities to be constructed.

We give first some references discussing the physics case for RIB

7.27 Los Alamos Working Group on the use of RIB for Nuclear Astrophysics (1990) April 10–12, Los Alamos
7.28 ISL – The Isospin Laboratory: Research Opportunities with Radioactive Ion Beams (1991) LALP-91-51
7.29 Garrett, J.D., Olsen, K.D. (eds.) (1991) A Proposal for Physics with Exotic Beams at the HHIRF
7.30 Butler, P.A. (1991) Spectroscopic Tools for the 90s: EUROGAM, SUSAN and RIB facilities: Nuclear and Atomic Physics with Accelerators of the Nineties, p. 209
7.31 Nuclear Structure and Nuclear Astrophysics: A Proposal (1992) University of Tennessee, Dept. Phys. UT-Th9201
7.32 The UNIRIB Consortium: A white paper (1995) Oak-Ridge
7.33 Vervier, J. (1995) Nucl. Phys. **A583**, 717
7.34 Nazarewicz, W., Sherrill, B., Tanihata, I., Van Duppen, P. (1996) Nucl. Phys. News **6**, 17

We also give references on how to make those RIB and include here a number of discussions about facilities, even though in the above references, specific facilities are discussed when presenting the physics case.

7.35 Ravn, H.L. (1991) RIB based on ISOL postacceleration, CERN-PPE/91-173
7.36 Garrett, J.D. et al. (1992) The Oak-Ridge RIB facility, ORNL Preprint
7.37 NUPECC Report (1993) European Radioactive Beam Facilities
7.38 Mueller, A.C. (1995) GANIL, IPNO-DRE 95-21
7.39 Villari, A.C.C. et al. (1995) RIB at Spiral, Nucl. Phys. **A588**, 267c
7.40 Lubkin, G.B. (1996) Oak-Ridge RIB facility, Physics Today, January, 24

Recent developments relating to the Isospin Laboratory ISL can be found in the ISL Newsletter (e-mail: rick@riviera.physics.yale.edu)

Finally, we give references to recent experiments in which radioactive ion beams of various types have been used and others that are being constructed.

7.41 Blackmon, J.C. et al. (1995) Phys. Rev. Lett. **74**, 2642
7.42 Decrock, P. et al. (1991) Phys. Rev. Lett. **67**, 800
7.43 Decrock, P. et al. (1993) Phys. Lett. **B304**, 50
7.44 Decrock, P. et al. (1993) Phys. Rev. **C48**, 2057
7.45 Galster, W. et al. (1991) Phys. Rev. **C44**, 2776
7.46 Michotte, C. et al. (1996) Phys. Lett. **B381**, 402
7.47 Motobayashi, T. et al. (1991) Phys. Lett. **B264**, 259

7.48 Ouellet, J.M.L. et al. (1992) Phys. Rev. Lett. **69**, 1896
7.49 Page, R.D. et al. (1994) Phys. Rev. Lett. **73**, 3066
7.50 Smith, M.S. et al. (1993) Phys. Rev. **C47**, 2740
7.51 Vermeeren, L. et al. (1994) Phys. Rev. Lett. **73**, 1935

And, finally, a short, technical review:

7.52 Huyse, M. et al. (1995) Nucl. Phys. **A588**, 313c

8. From Nucleons to the Atomic Nucleus: A Short Story

8.1 The Short Story

In this final chapter, having made a long journey through the exciting sub-atomic world where nucleons join together under the action of the strong force, we will try to view some of the major organizing elements that give the nucleus such a rich structure. At the same time, we emphasize and illustrate this theme with a number of brief examples. They demonstrate how the story of discovering the various basic features in the atomic nucleus is due, to a great extent, to the constant inventivity of experimentalists. It becomes clear that, whenever new technology has come within reach for probing the internal motion of nucleons in the nucleus, the nucleus has, time after time, shown new features and further elementary modes of nucleon motion. This line of experimental work has been connected largely to the development of new accelerators and detectors (Fig. 8.1), for observing nuclear interaction processes. This combination has led to a better understanding of how nucleons feel the nucleon–nucleon interaction and thereby build up the nuclear many-body system.

We have journeyed through the various distance and energy scales: observing the processes that play a role on the level of the nucleon itself and its correlations with other nearby nucleons; studying the various substructures that are formed inside the nucleus like clustering modes of nucleons; and, finally, concentrating on the large-scale mean-field properties of the nucleus which give rise to a largely independent-particle picture of nucleons moving in the average field generated by all other nucleons. We have also considered the various possible interactions with external fields: Electromagnetic probing of the internal motion of the nucleon has shown, and still is revealing, new aspects of how a nucleon behaves in the presence of A nucleons. Weak interactions that transform the nucleus into a most interesting laboratory, are able to test some of the basic symmetries governing the global interactions and invariance properties. We have also concentrated on what may happen, as a result of the strong force, when two "chunks" of nuclear matter (two large atomic nuclei) are smashed into one another using heavy-ion accelerator methods. The behavior of nuclear matter under these extreme conditions of high temperature and high density over a relatively large spatial volume may give rise to totally new phenomena, such as a possible phase transition

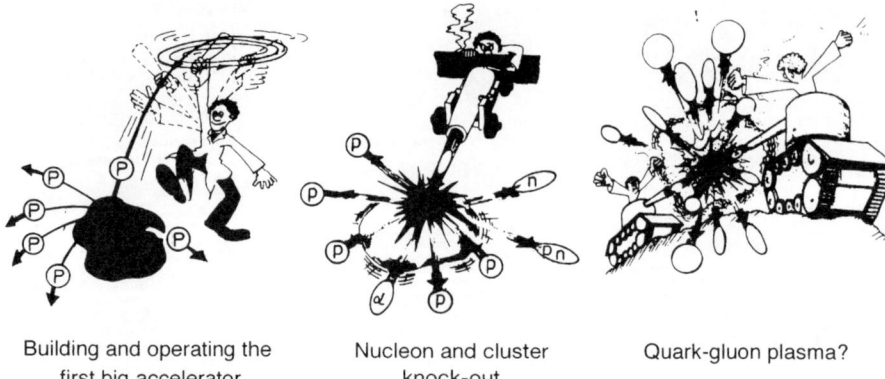

Building and operating the   Nucleon and cluster   Quark-gluon plasma?
     first big accelerator   knock-out

Fig. 8.1. Bigger and bigger accelerators. We illustrate how physicists have come to explore more and finer details using ever-increasing energies. With the lower energy (a few tens of MeV) cyclotrons, Van de Graaff, and electron accelerators, global properties of the nucleus could be observed via scattering of protons and/or electrons. This gives access to overall nuclear shape properties (*left*). With increasing accelerator energy (1–2 GeV), fine details of how protons and neutrons behave and move inside the nucleus have become accessible. Here, the scattered electrons can be detected together with nucleons or clusters of nucleons knocked out of the nucleus in the scattering process (*center*). On the very high energy side (a few GeV/A), the ultimate accelerators may show glimpses of the quark–gluon world (plasma state?) in some detail (*right*). (Adapted from Fig. 1.8 in F. Close, M. Marten, C. Sutton *The Particle Explosion* ©1994 Oxford University Press, with permission)

of regular nuclear matter made of nucleons into a phase where color confinement disappears and quarks and gluons are released in a plasma phase. This new phase might mimic some of the conditions that pertained in the very early phases of the universe. In this way we have seen that nuclear physics and astrophysics have a thin dividing line and that experiments under controlled laboratory conditions give access to processes that explain the present-day abundances of elements in the universe and in our solar system. With the advent of radioactive ion beam accelerator facilities, new directions and technical possibilities are opening up and will become accessible to nuclear physicists in the years to come.

The 65 year period of evolution in nuclear physics, originating around 1932 with the suggestion of a two-fold symmetry classifying the observed proton and neutron in a unified way, up to present-day symmetry concepts and microscopic attempts to understand the atomic nucleus, its internal structure, and its interaction with external fields, has been explosive and is not yet over.

The nucleus is not just a collection of nucleons and simply knowing some properties of nucleons and their interaction is quite insufficient to explain the observed nuclear phenomena. This is clearly demonstrated by the new layers of structure that have successively come onto the observational horizon, often quite unexpectedly.

This constant sequence of surprises, generated by intensive work in many laboratories worldwide, has provided the road to a deeper understanding and has set the stage for, and determined the trends in present day nuclear physics.

8.2 Nuclear Physics as a Piece of the Global Physics Jigsaw

Nuclear physics is a field that is strongly connected to and intertwined with other scientific disciplines (Fig. 8.2).

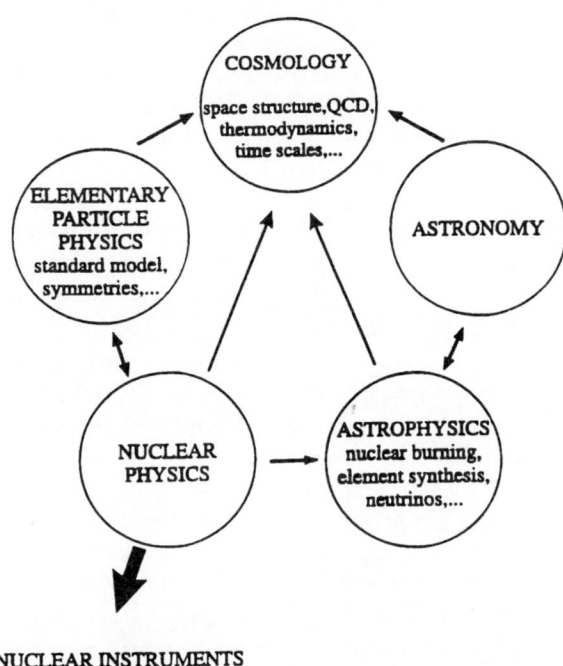

Fig. 8.2. Schematic illustration of the various interconnections between nuclear physics and a number of other fields of physics research. A line towards the more applied domains of nuclear instrumentation and methods is also presented. (Reprinted from A. Richter (1993) Nucl. Phys. **A553**, 417c. Elsevier Science, NL, with kind permission)

It shares a long boundary with the domain of astronomy and astrophysics because many of the basic processes fueling the stars in our universe are ruled by basic nuclear physics. They govern the large chain of nuclear reactions that keep our Sun and many other stars "alive", and result in the element distribution that we now observe in our solar environment. This is particularly the domain of the strong forces that build up nuclei and, through a number of capture processes, form all of the heavy elements of nature. The weak force is also very important in understanding nuclear astrophysics processes because, at the end of the regular burning phase, neutrinos are released in large numbers when protons are transformed into neutrons through the electron capture process. The subsequent interaction of neutrinos with nuclei largely determines what may happen in supernova processes. At present, these neutrino reaction processes are one of the major topics in nuclear physics, but now under laboratory conditions.

The separate fields of nuclear and particle physics have recently seen the emergence of such common ground. Here, the term 'intermediate energy' physics has been coined to characterize research in which the pure nucleonic degrees of freedom alone are no longer able to fully describe the observations. In the other direction, specific characteristics of mesons, and even deeper, quark and gluon variables also enter the picture. A word of caution should be given at this stage: One does not need the full power of quantum chromodynamics (QCD) to describe phenomena like superdeformation or shape coexistence in nuclei. There appears to be a 'filtering' of information (call this scaling or, better still, renormalization) from the deeper level (deeper in the sense of smaller length and higher energy scales) into the upper level (the level where nucleon degrees of freedom are the essential variables of the theory). In a schematic way, the parameters determining the electroweak and strong Lagrangian of particle physics do have to be averaged into a description where nucleon variables are dominant – called quantum hadrodynamics (QHD) – entering the region where nuclear structure properties play a role on a more global level (mean-field concept, global properties of the nucleus, and the nuclear interaction acting inside the nuclear medium).

In a different direction, developments in basic nuclear physics strongly linked to applied fields, for example through the construction of accelerators that offer unique and specific beams of particles. So, it is not just the increase of energy and intensity that has allowed some of the major developments in nuclear physics: Dedicated efforts have often been at the start of new research directions and we shall discuss some particular efforts in the next section. Strongly related to the former point has been the development of specific detector systems in the attempt to solve some very difficult problems. The advent of the very high-resolution and highly efficient gamma arrays operating over an almost full 4π geometry has given access to features hitherto impossible to detect, such as the gamma transitions in superdeformed bands and the connecting gamma transitions into the regular low-lying levels. Detector technology has also seen advances in detecting and in discriminating various species of particles and determining energies over a large interval. Further influence comes from the still growing computer facilities and computer performance.

On a more general level and supported by large-scale theoretical undertakings, a number of unifying themes, common to the atomic nucleus and many other quantum many-body systems, have been shown to exist. They are studied using both elegant general ideas of theoretical many-body physics and high-power numerical quantum Monte Carlo simulations of the nucleus.

8.3 The Importance of Technical Progress

Here we shall see that, with every new experimental technique for accelerating or detecting particles, the rich field of nuclear physics has revealed new, and most often unexpected features. We shall illustrate this by just a few selected examples although it could itself be the basis for discussing all developments in nuclear physics over the last 60 years.

Our information about the atomic nucleus comes via the detection of particles and gamma radiation emitted from the nucleus as it de-excites into its ground state. The early detectors for gamma radiation using NaI(Tl) detectors had rather poor resolution compared to the present-day Ge detectors. The latter detectors, put into arrangements that allow the Compton scattered processes to be added to the main photopeak process have opened totally new avenues in gamma spectroscopy. Added up in large arrays (Chap. 3) with a full 4π geometry, this has resulted in the discovery of the physics of superdeformed bands and also of the way heated nuclei behave when they cool off via the emission of a number of photons.

Accelerators for various particles have evolved far from the original Van de Graaff and cyclotron machines into very sophisticated equipment. One region where particular progress has been made is the field of electron accelerators in which a large spectrum of energies is now available. Lower energy accelerators (a few tens of MeV) and intermediate energy apparatus (100–200 MeV) are mainly used in a very specific way in order to study elementary nuclear modes of motion. On the higher energy side one now has a number of approximately 1 GeV accelerators of a continuous wave type (CW) delivering electrons continuously by the use of recirculation (race-track) techniques, sometimes combined with superconducting technology to keep magnets to manageable sizes. In Chap. 4, we discussed a number of such accelerators and the interesting physics that they can and will potentially reveal. At present, CEBAF represents the state-of-the-art with a 4-GeV electron final energy; it has just become operational. On the lower energy side, by concentrating on specific characteristics, there is still much physics to be studied and new discoveries are likely to be made. We refer to the 180° electron scattering facility at the Technische Hochschule, Darmstadt where, through magnetic dipole excitations of the nucleus (M1 excitations), a new mode of motion was discovered. Thereby, protons and neutrons are able, within deformed nuclei, to move in a way known as "scissors" motion.

In obtaining information about the atomic nucleus, a characteristic that is particularly important is the speed with which the excited levels decay and so, through the nuclear lifetime, give access to the nuclear dynamics. Most standard techniques of measuring lifetimes, using delayed coincidence methods between two subsequent photons feeding and de-exciting a given level, have a lower limit in the region of nanoseconds (ns). A new technique has been developed, largely at the high-flux reactor at Grenoble (ILL) where, using the neutron capture process, gamma rays are emitted by the nucleus as it seeks out its way into the ground state. Now, every time a photon is emitted, through conservation of momentum, the nucleus gets little kicks and so photons are emitted not from a nucleus at rest but from a moving nucleus. The photon is therefore doppler shifted and broadened and, by studying the precise shape of the emitted gamma ray in detail, one can learn about the full history of how the nucleus de-excites and about the half-lives of the nuclear level. This technique, called GRID (Gamma-Ray Induced Doppler broadening method), has opened a new "window" of nuclear lifetimes that were previously inaccessible. These data then allow an extension of tests of our theoretical description of the possible forms of motion of nucleons inside the nucleus. In a number of cases, clear-cut assignments could be made using the GRID method leading to a better description of atomic nuclei.

The advent of highly precise ISOL (isotope separator on line) systems, allowing nuclear physicists to create, transport, detect, measure, and analyze the properties of nuclei very far away from the region of beta-stable nuclei within the time of milliseconds, and even beyond, has greatly extended our knowledge about the way nuclei behave, not only near stability, but also when approaching the extreme edges of stability. This in itself is already a unique accomplishment achieved by a number of accelerators connected to ISOL systems. ISOLDE at CERN has been particularly successful over a long period and has thus acquired great expertise in the domain of studying exotic nuclei and disentangling their most secret properties. Efforts by physicists from the University of Mainz (Otten and co-workers) have made it possible to study the way in which electrons are coupled to the nucleus by looking at the hyperfine interactions via extensive laser investigations of the optical properties of these exotic ions. The tiny hyperfine interactions carry much information about intrinsic nuclear properties such as magnetism (nuclear dipole moments, nuclear quadrupole moments, nuclear charge radii, etc.). In unravelling these bits of information many surprises have emerged. The liquid drop concept for describing global properties of the nucleus predicts that with increasing A the nuclear size gradually increases (like $A^{1/3}$ for the radius). It was a big surprise to find that in a number of mass regions, this trend was not just modified but, occasionally, totally reversed. The Au, Tl, Pb region is a most instructive and dramatic exhibition of this. It illustrates that one may still encounter the unexpected in the study of the atomic nucleus (Fig. 8.3).

In studying very light nuclei in the He, Li, Be region (see Chap. 3) very loosely bound quantum systems were discovered, possessing almost unbound nucleons moving outside a rather bound core system. The nucleus ^{11}Li is one of the best examples of these halo systems which are, at present, not very well understood. They may form a special kind of configuration of the nucleus along the path of rearranging protons and neutrons before the system becomes totally unbound. Much research remains to be performed in this field of exotic nuclear physics.

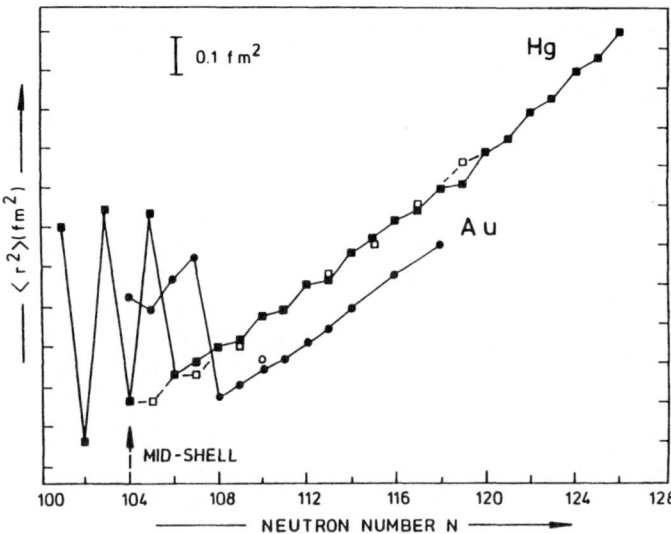

Fig. 8.3. Variation in the nuclear in the high mass region (the Pb region of nuclei). Just below the $N = 126$ neutron closed shell, a liquid drop model variation is followed rather closely. Near mid-shell ($N = 104 - 108$), however, very large changes show up in the experimental data that cannot be understood by regarding the nucleus as a charged liquid drop

We saw in Chap. 7 that new capabilities are coming of age in accelerating radioactive ion beams, and that new technical possibilities are becoming available to explore a region of atomic nuclei that was previously held to be beyond the reach of laboratory study. We discussed the facilities and their capabilities in quite some detail in Chap. 7. The significance for nuclear physics is twofold: (i) it provides ways of exploring nuclei much further away from the region of beta stability and, (ii) it creates unique ways to study reactions that occur at the "heart" of stars in the formation of elements and are thus basic to our understanding of why our universe, and in particular our solar system is the way it is. Further new discoveries will certainly follow in this still very young field of nuclear astrophysics.

This is a rather incomplete list of illustrations of how new technical developments have ignited new directions of research; many others could be mentioned. As a last point, I would like to refer to the basic studies that have become possible thanks to the ability to collect slow and cold neutrons at reactor facilities (neutron bottles, etc.). Here a number of very fundamental questions have come within reach: measurement of the precise half-life of the neutron has put constraints on the standard model and on the constitution of the early universe. Neutrons hold a number of secrets to our understanding of nature. The search for a possible electric dipole moment is still in progress.

This quantity is a very significant parameter for going into physics beyond the standard model.

8.4 How Well Do We Understand the Atomic Nucleus?

At present it is not possible to perform ab initio calculations that generate the properties of the atomic nucleus starting from a given nucleon–nucleon interaction. Attempts have been undertaken to describe nucleon motion in the simplest few body systems like ^{2}H, ^{3}He, ^{4}He using Monte Carlo methods. They are found to agree with the main assumption of some kind of single-particle motion and orbitals in which the individual nucleons are moving. Present methods which make maximum use of the computer power available today are large-scale shell model studies in which an "effective" force (the pure nucleon–nucleon force corrected for medium effects caused by the presence of $A - 1$ nucleons) is diagonalized in a full basis. At present, the full sd shell and part of the fp shell can be treated with model spaces extended to include up to $10^8 - 10^9$ independent configurations. There is a clear limit to this method, the moreso because the truncation methods used in the big shell model studies can miss a number of low-lying excitations actually observed in nuclei. One needs truncation schemes that are better adapted to the dominant degrees of freedom exhibited in the atomic nucleus. A very promising approach combines a Monte Carlo method to determine the optimal basis consistent with the main degrees of freedom and conserved symmetries like rotational invariance, isospin invariance, etc. Then, in a second step, the interaction is diagonalized within the chosen Monte Carlo model space (called QMCD). At present, calculations within a full fp shell are being carried out and we show, in Fig. 8.4, an illustration for ^{64}Ge. This method developed by Otsuka and co-workers (describing detailed nuclear spectroscopic information) is in contrast to the former Shell Model Monte-Carlo method (SMMC), worked out by Koonin and co-workers which enables the description of thermal properties of atomic nuclei (see Chap. 3 for references).

Apart from the approach that treats the nuclear many-body system in terms of the individual nucleon degrees of freedom interacting through an effective two-body interaction, one can also attempt to understand the organization of nucleons inside the atomic nucleus using symmetry concepts. One of the early applications of continuous symmetries in nuclear physics consisted in treating the charge character of the individual nucleons (see also Chap. 3) within the SU(2) group. Combining this with the intrinsic fermion spin character of protons and neutrons led to the Wigner supermultiplet SU(4) symmetry. In later years, it was shown by Elliott that the nuclear shell model can be used to classify modes of quadrupole collective motion thereby bridging the gap between the microscopic and macroscopic approaches to nuclear structure. In recent decades, combining the pairing (short-range) and

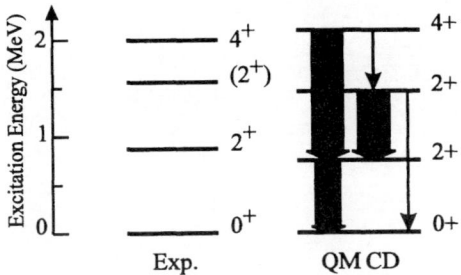

Fig. 8.4. State-of-the-art calculations using the nuclear shell-model techniques in order to probe the nuclear many-body system. Quantum Monte Carlo (QMC) methods have been extensively used in recent years (Chap. 3). The present example, the nucleus ^{64}Ge, illustrates the capabilities of these QMC methods, exploiting the limits of present-day computing power. (Taken from M. Honma et al. (1996) Phys. Rev. Lett. **77**, 3315. American Physical Society, with permission)

quadrupole (long-range) components within a boson model symmetry, a U(6) group structure has been proposed by Arima and Iachello. It resulted in a largely unified approach encompassing a description of diverse excitations in the atomic nucleus. In all of these symmetry classifications, use has been made of nucleonic excitations outside of closed shells. Extensions to symmetry groups in which both the horizontal (excitations within a given major shell) and vertical (excitations across closed shells forming many-particle many-hole configurations) ways of moving nucleons around inside the nucleus are combined will require the study of larger groups and may indeed lead to an elegant description of the atomic nucleus.

Exciting the atomic nucleus up to a high energy or, equivalently, large intrinsic temperature, often drives the system into a regime where symmetries existing at low energy cease to hold. One can have confidence in the shell model for describing nuclei at high excitation energy, but this only for light nuclei. In the heavier nuclei and even in light nuclei, when one considers all nucleons without the constraint of closed shells forming an inert core, and forms all states via combinatorial methods or, uses statistical methods, the situation becomes too complex to be handled using standard shell model methods. Here, the main quantities of interest are no longer the energy positions of all individual levels but rather, level densities, thermodynamical properties, dissipation of angular momentum in rotational damping processes, and properties related to the quantum statistical behavior of a collection of many nucleons. One must also ask how far the symmetries determining nuclear structure properties at low energy extend into the region of high energy.

Various energy regions, exhibiting different major excitation modes, are schematically illustrated in Fig. 8.5

Our understanding of how the atomic nucleus interacts with external fields has also progressed significantly. Electromagnetic interaction processes

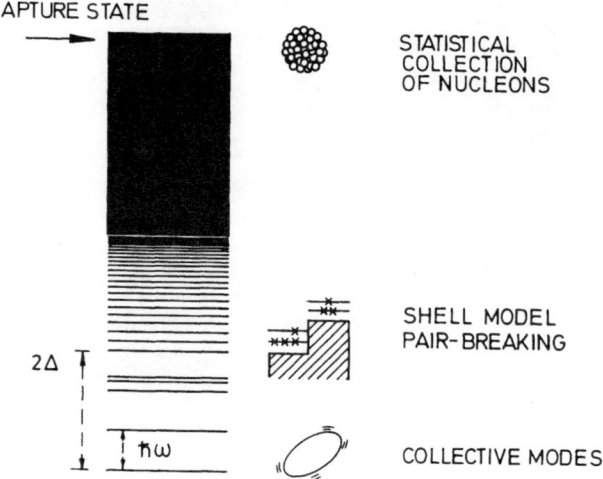

Fig. 8.5. Various regions where the nuclear many-body system exhibits quite different basic modes of motion: At the lower energies, collective modes of motion (vibrations, rotations, etc.) occur. At around the energy needed to break the nucleon–nucleon pair correlations (about $2\Delta \simeq 2$ MeV), typical shell model correlations dominate the excitation modes in the atomic nucleus. At high energies, ($\simeq 8$ MeV at the nucleon separation energy), statistical properties become dominant. Occasionally, simple modes of motion may survive at such high energies e.g. superdeformed band structures, giant multipole resonances, etc.

(Chap. 4) have yielded a detailed understanding of how nucleons move near the Fermi level in the atomic nucleus, of how the interaction process is able to knock out a nucleon or clusters of nucleons from the nucleus, and have been used to study nucleon–nucleon short-range correlations. A number of review papers and recent books describe this field in great detail and outline the present boundaries of our knowledge (see References in Chap. 4). The weak interaction processes have not been studied in so much detail as yet; in recent years, however, the field of neutrino interactions with nuclei has become very active, despite the very small cross-section of a neutrino interacting with a single nucleon. This is compensated by the very intense neutrino sources at various accelerators. In the coming years, a number of decisive neutrino experiments are planned that should shed light on neutrino–nucleus interactions, on the long-standing solar neutrino problem, and on the related topic of a possible non-vanishing neutrino mass.

When studying nuclear physics, one may wonder where relativity should come in and for which properties it plays a dominant role. Efforts over a number of years by Ring and co-workers using, besides the nucleon mean field properties, meson fields that carry the interaction and constrain the dynamics to be consistent with a relativistic treatment, have produced impressive results. A unique identification of the major properties whose description requires such relativistic models remains to be made at low excitation energy. A number of efforts and ideas are being pursued in this direction. Work attempting to connect a pseudo-spin symmetry observed in the mean-field motion of individual nucleons in both spherical and deformed nuclei to an underlying spin–orbital relativistic interaction structure is in progress (Ginocchio et al.).

It is difficult to find an appropriate conclusion. Suffice to say that impressive progress has been made, from the early concepts of protons and neutrons moving in a simple spherical or deformed average potential, towards full-scale symmetry-dictated shell model studies and the study of more complex group structures encompassing a large family of nuclear modes of motion, extending up to high energy. New techniques and developments will surely help us to better understand and find underlying simplicity in the highly varied types of nuclear excitations. And just as surely, the new results will bring with them new challenges.

8.5 Further Reading

In previous chapters we have given extensive reference to a wide variety of literature ranging from popular accounts to specialist textbooks and technical papers. Here we present only a short set of references. These are reports looking into longe-range planning and coordination of experimental and theoretical research in nuclear physics. These reports emphasize the importance of the research carried out on the level of smaller groups at university laboratories and research institutes. We also include a few relevant papers that were presented at some of the recent nuclear physics conferences.

8.1 The DOE/NSF Nuclear Science Advisory Committee (1996) Nuclear Science: A Long Range Plan

8.2 NUPECC Report (1991) Nuclear Physics in Europe, Opportunities and Perspectives

8.3 NUPECC Report (1993) European Radioactive Beam Facilities

8.4 ISL Report (1991) The Isospin Laboratory: Research Opportunities with Radioactive Nuclear Beams, LALP 91-51

8.5 Richter, A. (1993) Trends in Nuclear Physics, Nucl. Phys. **A553**, 417c

8.6 Detraz, C. (1995) Perspectives in Nuclear Physics, Nucl. Phys. **A583**, 3

8.7 Feshbach, H. (1995) Closing Remarks, presented at the Conference "Nucleus–Nucleus Collisions V" Nucl. Phys. **A583**, 871

8.8 Siemssen, R. (1995) Concluding Remarks, presented at "Int. Symp. on the Physics of Unstable Nuclei, Niigata" Nucl. Phys. **A588**, 371c

8.9 van der Woude, A., (1995) Impact and Application of Nuclear Science Opportunities and Perspectives, Nucl. Phys. **A583**, 51

8.10 Koonin, S.E. (1994) Perspectives on Nuclear Structure, Nucl. Phys. **A574**, 1c

Glossary of Acronyms

AAPPSB	Assoc. of Asia Pacific Phys. Soc. Bulletin
AECL	Atomic Energy of Canada Limited
AGS	Alternating Gradient Synchrotron
AMPS	Amsterdam Pulse Stretcher
ANTOINE	Strasbourg shell model code (E. Caurier, 1989)
ARENAS 3	Acceleration of Radioactive Elements for Nuclear, Astrophysics and Solid State Sciences
ATLAS	Argonne Tandem/Linac Accelerator System
BATES	linear accelerator center at Massachusetts Institute of Technology
BGO	bismuth germanate oxide
BNL	Brookhaven National Laboratory
BRAHMS	smaller detector system at RHIC
BUU	Boltzmann-Uehling-Uhlenbeck
CEBAF	Continuous Electron Beam Accelerator Facility
CELSIUS	Cooling with Electrons and Storing of Ions from the Uppsala Synchrocyclotron
CERN	Conseil Européen de Recherche Nucléaire
CIME	Cyclotron d'Ions à Moyenne Energie (Medium Energy Ions Cyclotron)
CNO-cycle	carbon–nitrogen–oxygen cycle
CNO	carbon–nitrogen–oxygen
COSY	Cooler Synchrotron Jülich
CRN	Centre de Recherches Nucléaires (de Strasbourg)
CYCLONE	Cyclotron of Louvain-la-Neuve
DAΦNE	Double Annular Φ-Factory for Nice Experiments
DESY	Deutsches Electronen Synchrotron
DOE/NSF	Department of Energy/National Science Foundation
ECR	electron cyclotron resonance
ECU	European Currency Unit (in August 1997, 1 ECU was worth US$ 1.06)
EEF	European Electron Facility
ELFE	Electron Laboratory for Europe
EMC	European Muon Collaboration

EOS	equation of state
EUROBALL	European Gamma Array Project (collaborators: Denmark, France, Germany, Italy, Sweden, UK)
EUROGAM	European Gamma Array (collaborators: UK, France)
FEL	free electron laser
GALLEX	Gallium Experiment (at Gran Sasso, Italy)
GAMMASPHERE	Gamma Array (in 4π geometry) USA
GANIL	Grand Accélérateur National d'Ions Lourds
GASP	Gamma Ray Spectrometer (Legnaro, Padova)
GDR	Giant Electric Dipole Resonance
GDR-IS	Giant Electric Dipole Resonance – Isoscalar
GOE	Gaussian-orthogonal ensemble of matrices (or interactions)
GRID	gamma-ray induced doppler broadening method
GSI–LBL	Gesellschaft fur Schwerionenforschung – Lawrence Berkeley Laboratory
GSI	Gesellschaft fur Schwerionenforschung (Darmstadt)
GUT	grand unified theory
HERA	Hadron Electron Ring Anlage (Hamburg)
HERMES	Hera measurements of spin dependent structure functions (at HERA)
HHIRF	Holifield Heavy-Ion Research Facility
HRIBF	Holifield Radioactive Ion Beam Facility
IBM	interacting boson model
ILL	Institut Laue-Langevin (Grenoble)
IMF	intermediate mass fragment
IPM	independent particle model
IPN	Institut de Physique Nucléaire at Orsay
IPNO-DRE	Institut de Physique Nucléaire at Orsay
ISAC	Isotope Separator and Accelerator (TRIUMF upgrade for accelerated radioactive beams)
ISIS	the name given to the intense neutron spallation source at Rutherford Appleton Laboratory (RAL)
ISL	Isospin Laboratory
ISN	Institut de Sciences Nucléaires (Grenoble)
ISOL	Isotope Separator On-Line
ISOLDE-3	Isotope Separator On-Line (upgrade at the Proton Synchrotron Booster, CERN)
ISOLDE	Isotope Separator On-Line (CERN)
JETP	Journal of Experimental and Theoretical Physics
JHP	Japanese Hadron Project
JINR	Joint Insitute for Nuclear Research (Dubna)
KEK	National Laboratory for High-Energy Physics at Tsukuba
LALP	Los Alamos Laboratory Preprint

LAMPF	Los Alamos Meson Physics Facility
LBL	Lawrence Berkeley Laboratory
LEAR	Low-Energy Antiproton Ring
LEP	Large Electron–Positron Collider
LHC	Large Hadron Collider
LLNL	Lawrence Livermore National Laboratory
LSND	Liquid Scintillator Neutrino Detector (Los Alamos)
MAMI	Mainz Microtron
MIT	Massachusetts Institute of Technology
MSU	Michigan State University
MSW	Mikheyev, Smirnov, and Wolfenstein
NDE	Nuclear Data Ensemble
NIKHEF	Nationaal Instituut voor Kernfysica en Hoge-energiefysica
NSCL	National SuperConducting Laboratory
NUPECC	Nuclear Physics European Collaboration Committee
NUPP	Nuclear and Particle Physics
OPEP	one-pion exchange potential
ORNL	Oak-Ridge National Laboratory
OXBASH MSU	Oxford-Buenos-Aires Shell Model Code (at MSU)
PHENIX	major detector system set-up at RHIC
PHOBOS	smaller detector system set-up at RHIC
PIAFE	Projet d'Ionisation et d'Accélération de Faisceaux Exotiques (Grenoble)
PLASTIC	Spherical shell containing 800 plastic detectors for heavy-ion reaction studies (GSI-LBL collaboration)
PNC	parity non-conservation
PRIMA	Isolde post accelerator for radioactive ions (CERN)
PSI	Paul Scherrer Institute (at Villigen, Switzerland)
PWIA	plane wave impulse approximation
QCD	quantumchromodynamics
QHD	quantumhadrodynamics
QMC	quantum Monte Carlo
QMCD	quantum Monte Carlo diagonalization
REX-ISOLDE	Radioactive Beam Experiments at ISOLDE
RFQ	radio-frequency quadrupole
RHIC	Relativistic Heavy-Ion Collider (at Brookhaven National Laboratory)
RIB	radioactive ion beams
RIKEN	Institute of Physical and Chemical Research (Japan)
RITSCHIL	Utrecht shell model code
RPA	Random-Phase Approximation
SATURNE	Synchrotron at the French National Laboratory for Hadron Physics (Saclay, to be closed by the end of 1997)
SLAC	Stanford Linear Accelerator Center

SMC	Spin Muon Collaboraton
SMMC	shell model Monte Carlo method
SNO	Sudbury Neutrino Observatory
SNU	Solar Neutrino Unit
SPIRAL	Système de Production d'Ions Radioactifs et d'Accélération en Ligne (Production and On-Line Acceleration System for Radioactive Ions)
SPS	Super Proton Synchrotron
SSC2	Separated Sector Cyclotron Number 2
STAR	major detector system set-up at RHIC
SURA	South-Eastern Universities Research Association
SUSAN	Spectrometer for Universal Selection of Atomic Nuclei (P. Butler, 1991)
TESSA-3	Total Energy Suppression Shield Array 3 (Daresbury)
TISOL	Thick-Target Isotope Separator On-Line (TRIUMF)
TJNAF	Thomas Jefferson National Accelerator Facility
TRIUMF	Tri-Universities Meson Physics Facility (Vancouver, Canada)
TUNL	Triangle University Nuclear Laboratory
UNIRIB	University Radioactive Ion Beams Consortium
VAMPIR	shell-model code from University of Tübingen

Index

Springer
and the
environment

At Springer we firmly believe that an international science publisher has a special obligation to the environment, and our corporate policies consistently reflect this conviction.
We also expect our business partners – paper mills, printers, packaging manufacturers, etc. – to commit themselves to using materials and production processes that do not harm the environment. The paper in this book is made from low- or no-chlorine pulp and is acid free, in conformance with international standards for paper permanency.

 Springer

Printing: Mercedesdruck, Berlin
Binding: Buchbinderei Lüderitz & Bauer, Berlin

Key to the Nuclear Mass Table

Magic nucleon numbers lie between double lines

Stable isotope : black

Unstable isotope : purple = β^+ emitter and electron capture
  : blue = β^- emitter
  : yellow = α emitter
  : orange = p emitter
  : green = nucleus undergoing spontaneous fission

Boundary lines : red-dashed = proton binding energies $B_p = 0$
  : blue-dashed = neutron binding energies $B_n = 0$
  : blue-dashed = fission barrier $B_f < 4$ MeV

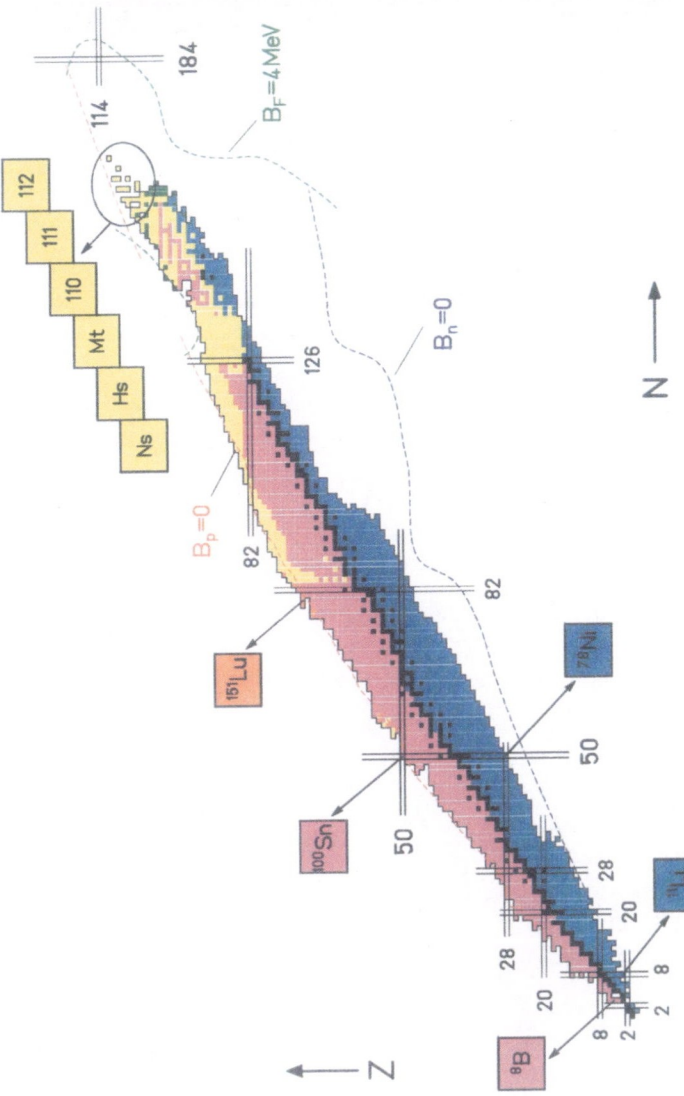